Architectural Aerodynamics

ARCHITECTURAL SCIENCE SERIES

Editor

HENRY J. COWAN

Professor of Architectural Science
University of Sydney

Previously published

Thermal Performance of Buildings
by J. F. VAN STRAATEN

Fundamental Foundations
by W. FISHER CASSIE

Models in Architecture
by H. J. COWAN, J. S. GERO, G. D. DING, and R. W. MUNCEY

Principles of Natural Lighting
by J. A. LYNES

Electrical Services in Buildings
by P. JAY and J. HEMSLEY

Architectural Acoustics
by ANITA LAWRENCE

Fire and Buildings
by T. T. LIE

Psychology for Architects
by D. CANTER

Spatial Synthesis in Computer-Aided Building Design
edited by C. M. EASTMAN

Wind Loading on Buildings
by J. MACDONALD

Design of Building Frames
by J. S. GERO and H. J. COWAN

Building Services
by P. R. SMITH and W. G. JULIAN

Sound, Man and Building
by L. H. SCHAUDINISCHKY

Man, Climate and Architecture—2nd ed.
by B. GIVONI

Architectural Aerodynamics

by

R. M. AYNSLEY
*Department of Architectural Science,
University of Sydney, Sydney, Australia*

W. MELBOURNE
*Professor of Fluid Mechanics,
Monash University, Clayton, Victoria, Australia*

and

B. J. VICKERY
*Professor of Engineering Science,
University of Western Ontario, London, Canada*

APPLIED SCIENCE PUBLISHERS LTD
LONDON

APPLIED SCIENCE PUBLISHERS LTD
RIPPLE ROAD, BARKING, ESSEX, ENGLAND

ISBN: 0 85334 698 4

WITH 171 ILLUSTRATIONS

Composed by Eta Services (Typesetters) Ltd., Beccles, Suffolk
Printed in Great Britain by Galliard (Printers) Ltd., Great Yarmouth, Norfolk

Preface

The engineering profession has made significant progress in recent years in the ability to study and predict wind effects on buildings in wind tunnels. However, many people in the architectural, building and town planning professions are still uncertain as to how they can contribute, to minimise wind loads and environmental wind problems around buildings. Readers new to these fields often find articles and proceedings of conferences on these topics difficult to decipher because of their large amount of assumed knowledge in meteorology, fluid mechanics, wind tunnel modelling and dynamic structural behaviour.

This book attempts to explain, in simple terms, the basic theories and terminology used in the broad field of architectural aerodynamics and the tremendous influence that architects and property developers have on the general airflow around buildings, by determining the building's shape and its location relative to adjacent buildings.

As many environmental wind problems around buildings are due to the shape of buildings, location of openings or relative position of buildings, decision makers in the early stages of development, such as planners and architects, can benefit greatly from inexpensive wind tunnel studies at the preliminary design stage. Such studies can help designers identify serious wind problems at a stage when basic building concepts can be changed without prohibitive expense and avoid embarrassing and costly changes in the final design stage.

Portions of the text dealing with natural ventilation bring together the fluid mechanics of airflow with improved wind data, which allows assessment of probable ventilation for thermal comfort.

In general, the contents of this text should enable readers to proceed to more specialised publications in the field with more confidence, having an understanding of the basic principles involved in architectural aerodynamics.

R. M. AYNSLEY
W. H. MELBOURNE
B. J. VICKERY

Contents

Notation xi

Chapter 1. *Historical Developments in Architectural Aerodynamics* . 1

1.1. Definition 1
1.2. Early attempts to design for wind 1
1.2.1. Impact of the industrial revolution on town planning 9
1.3. Developments in instrumentation 11
1.3.1. Wind direction indicators 12
1.3.2. Atmospheric pressure measurement . . 12
1.3.3. Wind speed measurement 12
1.3.4. Wind force scales 17
1.3.5. Wind pressure measurement 17
1.4. Developments in fluid mechanics 25
1.5. Developments in meteorology 29
1.6. Simulated wind studies 30
1.6.1. Early wind tunnel studies of model buildings . 32
1.6.2. Introduction of thick boundary layers in wind tunnels 32
1.7. Full scale studies of wind effects 39
1.8. Effect of tall building development on architectural aerodynamics 39
1.9. Current directions in architectural aerodynamics . 41

Chapter 2. *Basic Aerodynamics* 42

2.1. Flow fields and regions 42
2.2. Streamlines and continuity 43
2.3. Bernoulli's equation and pressure fields . . . 45
2.4. The venturi meter and pitot-static tube . . . 48
2.5. Energy considerations in flow systems . . . 51
2.5.1. The energy equation 51
2.5.2. A duct system 53
2.5.3. Free ventilation 54
2.5.4. Stack effect 54
2.6. Resistance to fluid motion 56
2.6.1. Viscosity 56
2.6.2. Drag 58

2.7. Dimensional analysis, scaling and some important coefficients 58
2.8. The boundary layer and separation. 62
2.9. The effects of Reynolds number 65
2.10. The effects of turbulence 67
2.11. Wind tunnel testing 71
2.11.1. Types of wind tunnel testing 72
2.11.2. Natural wind models 73
2.11.3. Model measurements 76

Chapter 3. Physical and Statistical Properties of the Wind 79

3.1. Major storm types 79
3.2. Influence of height and terrain 85
3.2.1. The boundary layer 85
3.2.2. Mean speeds 88
3.2.3. Turbulence levels 92
3.2.4. Gust speeds 93
3.3. Wind speed spectra 94
3.3.1. Long term spectra 95
3.3.2. Turbulence spectra 96
3.3.3. Gust size 98
3.4. Topographic effects 98
3.5. Wind climate 101
3.5.1. Probability distributions 101
3.5.2. Extreme wind speeds 104

Chapter 4. Wind Loads and Structural Response 108

4.1. Introduction 108
4.2. Dynamics of structures 108
4.2.1. Spring–mass–dashpot systems 108
4.2.2. Free vibrations 110
4.2.3. Sinusoidal forcing 111
4.2.4. Random forcing 113
4.3. Design wind speed 116
4.3.1. Return periods 116
4.3.2. Standard or basic design speeds 116
4.3.3. Local design speeds 118
4.4. Design wind loads 119
4.4.1. Quasi-static methods 119
4.4.2. Gust factor methods 124
4.4.3. Wind tunnel methods 125
4.5. Wind load calculations 126
4.5.1. External pressure distribution 126
4.5.2. Internal pressures 133
4.5.3. Loads on walls and roofs 138
4.5.4. Local loads 140

4.6. Wake induced dynamic loads 141
4.6.1. Spectrum of across-wind forces . . . 141
4.6.2. Vortex shedding 142
4.6.3. Broad band excitation 143
4.7. Aerodynamic damping 143
4.8. Torsion 145
4.9. Design criteria 147
4.9.1. Structural failure 147
4.9.2. Cladding failures 148
4.9.3. Deflections 148
4.9.4. Accelerations 148

Chapter 5. *Environmental Wind Flows Around Buildings* . . . 150

5.1. The wind flow field around a building . . . 150
5.2. Environmental wind speed criteria 154
5.2.1. Wind conditions which cause people to be blown over 154
5.2.2. Acceptable wind conditions. . . . 155
5.2.3. Relationships between peak gust and mean wind speeds and frequency of occurrence . 159
5.3. The prediction of wind environment around a building 163
5.3.1. Wind tunnel testing techniques . . . 163
5.3.2. Example of a wind tunnel test . . . 166
5.4. Guide lines for avoiding high induced wind speeds at ground level. 173
5.5. Dispersal of gaseous pollutants 176

Chapter 6. *Natural Ventilation* 180

6.1. Ventilation rates for carbon dioxide and odour control 180
6.2. Ventilation to improve thermal comfort in hot humid environments 181
6.3. Energy sources for natural ventilation . . . 182
6.4. Applications of stack effect for natural ventilation . 184
6.4.1. Example 185
6.5. Applications of wind forces to natural ventilation . 188
6.5.1. Airflow through an opening due to a static pressure difference 189
6.5.2. Airflow through openings in series . . 192
6.5.3. Sources of pressure coefficient data for estimating natural ventilation 194
6.5.4. Sources of local wind speed data for estimating natural ventilation 195
6.5.5. Sources of discharge coefficients for estimating natural ventilation 202
6.5.6. Advantages and disadvantages of pressure difference method for estimating natural ventilation 210

6.6. Velocity coefficient method of estimating natural ventilation 217
6.6.1. Sources of velocity coefficients and typical applications 223
6.6.2. Advantages and disadvantages of velocity coefficient method 228
6.7. Design considerations for airflow patterns inside buildings 229
6.8. Flow visualisation techniques 232

Appendix. Dynamic Viscosity and Density of Air and Water as a Function of Temperature 237

References 238

Index 247

Notation

A cross-sectional area normal to the flow
A area of opening
A_{n/n_0} dynamic amplification factor
a acceleration
$\frac{1}{a}$ wind speed dispersion
b crosswind dimension
C coefficient determined by experiment
C_D drag coefficient
C_d discharge coefficient
C_L lift coefficient
C_M moment coefficient
C_p pressure coefficient
$\hat{C}_p$ peak pressure coefficient
$\bar{C}_p$ mean pressure coefficient
C_v velocity coefficient
d downwind dimension
E modulus of elasticity
F force
F Froude number
g gravitational acceleration
H average height difference between inlets and outlets in a building stack
h vertical height difference between manometer fluid levels
h building height
I_V longitudinal turbulence intensity
I_w vertical turbulence intensity
K spring stiffness
K Weibull constant
k surface drag coefficient
L dimensional notation for length
M dimensional notation for mass
M Mach number
n frequency
n_1 damped natural frequency
n_0 undamped natural frequency
P power
P probability function
p static pressure
p_l static pressure loss
p_g static pressure gain
Q mean volumetric flow rate
q dynamic pressure
R gas constant
R return period
Re Reynolds number
r radius
S Strouhal number
$S(n)$ spectrum of n
s separation distance
T time interval
T absolute temperature

t dimensional notation for time
U modal wind speed
u velocity component in the principal direction of flow
$\bar{u}$ mean velocity component in the principal direction of flow
u' deviation of the velocity component from the mean velocity in the principal direction of flow
V_z wind speed at height z above ground
V_G wind speed at gradient height
V_{10} mean wind speed 10 m above ground
$\hat{V}$ peak gust wind speed
$\bar{V}$ mean wind speed
W vertical velocity
z height above a horizontal reference datum

α mean wind speed profile exponent
β gust wind speed profile exponent
δ logarithmic decrement
$\hat{\delta}$ peak across wind deflection
ζ damping expressed as a fraction of critical damping
η efficiency
θ wind direction
λ wavelength
μ dynamic viscosity
ρ air density
ρ_b bulk density of building
ρ_M density of manometer fluid
σ_V standard deviation of longitudinal velocity component
$\sigma_V{}^2$ mean-square value of deviations of instantaneous wind speed from the mean wind speed
σ_ω standard deviation of vertical velocity component
τ shear stress
ϕ phase angle difference

Chapter 1

Historical Developments in Architectural Aerodynamics

1.1. Definition

Architectural aerodynamics is basically the relationships between wind and buildings. These relationships are generally regarded as including two main areas—structural and environmental. Structural considerations are wind loading and building motions due to wind. Treatment of wind loading and building response in this book is limited to considerations of interest to architects and builders and does not cater for the in-depth needs of structural designers. The principal environmental considerations are natural ventilation, pedestrian wind comfort in spaces around buildings, wind generated noise and the effect of wind on building services particularly discharge of smoke and fumes, and air conditioning intakes and exhaust ducts.

1.2. Early Attempts to Design for Wind

Although the term 'architectural aerodynamics' is relatively new, the concern for wind effects on and around buildings dates back to early civilisations. The design of many ancient towns was influenced by wind effects. Without the aid of wind tunnel techniques or instruments, early planners had to learn from the errors of past planning. Many keen observations of wind effects were recorded to assist early planners.

The influence of wind effects on planning has been observed in the remains of the town of Kahan, founded about 2000 B.C. under the 12th dynasty in ancient Egypt (Fig. 1.1). In layout it reveals divisions between the workmen's district in the west, with very small houses, and the 'notables' district in the north. The large houses of the 'notables' face the pleasant north wind, while those of the workmen in the west are exposed to the hot desert wind [1.1].

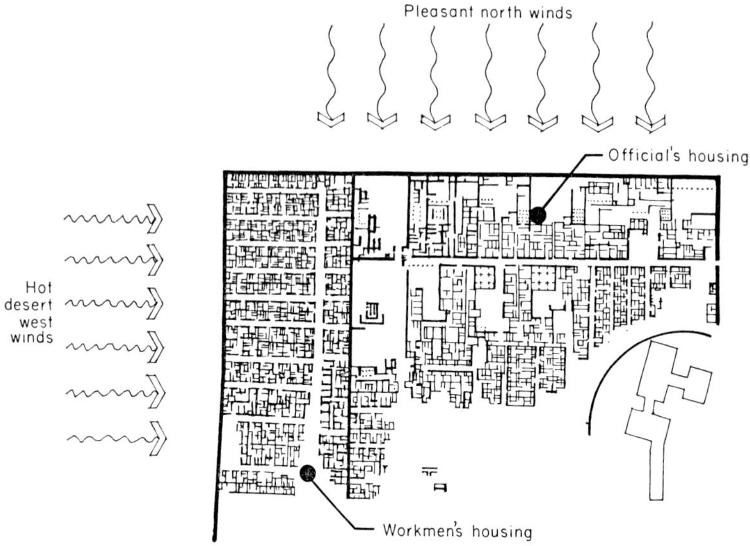

Fig. 1.1. Sketch of housing layout at Kahun, Egypt, around 2000 B.C.

Ancient Chinese dynasties after the Chou dynasty had principles for town development laid down in the 'Feng-shui' doctrine which treats the integration of human works into the pattern of the universe. (feng—wind, floods of heaven; shui—water, floods of earth) [1.2]. Rules governing the design of houses were concerned with spirits and forces within the house that could be controlled by arrangements to encourage the entry of air and light, and shelter from the wind and cold (Fig. 1.2).

Aristotle was the first to write in length about the mysterious winds in his 'Meteorologica' in the fourth century B.C. Although his attempts at explaining the origin of wind, based on the impossibility of a vacuum in nature, have since proved unacceptable, his town planning principles for Greek cities were more apt:

> 'The healthiest cities are those which stand upon a slope inclining to the east so that the winds blow from the quarter of the sun rise. The next best aspect is one that is sheltered from the north wind as cities so sheltered enjoy milder winters'.

Theophratus, a pupil of Aristotle, had fanciful ways of predicting the weather. He said a wind could be in the offing when a dog rolled on the ground, or ducks flapped their wings, or when there was a particularly bright display of shooting stars. Wind was attributed

Fig. 1.2. Sketch from a Chinese drawing showing mountains to the north, water to the south, governing the site for a town.

with the ability to impregnate mares and vultures, while a whirlwind would be created by sprinkling vinegar upon the wind [1.3]. More specifically if the feet swelled there would be a change to the south wind although this sometimes indicated a hurricane and similarly, if a man had a shooting pain in the right foot [1.4].

Another Greek, Xenophon, around 300 B.C., suggested building the southern side of houses higher than the northern in order to allow good sun penetration on porticos on the south wall and help keep out the cold north winds. However, around this time the Greeks became more occupied with the allocation of gods for the winds from various directions, such as Zephyros for the west wind. They devised a wind-rose whose points were named after the gods of those winds (Fig. 1.3).

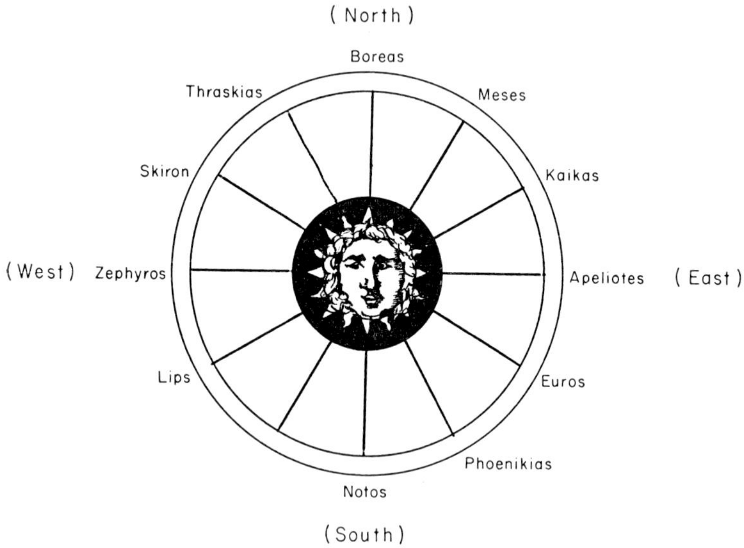

Fig. 1.3. The Greek wind-rose of Aristotle and Theophratus.

The Tower of the Winds erected in Athens during the second century B.C. has on each of its octagonal sides a sculptured motif depicting the characteristics of winds from eight points of the compass (Fig. 1.4). The Romans, who flourished from the third century B.C. to the first century A.D., made little effort towards explaining the origins of winds. They adopted the Greek knowledge of the winds and replaced the Greek names with Latin equivalents [1.3].

An exception was Vitruvius, a Roman architect and engineer of the first century B.C. He wrote his 'Ten Books on Architecture', the oldest and most influential book on architecture [1.5]. In this he referred to a number of considerations given to the design of buildings and towns in relation to wind. These included the selection of the site for the fortified town, the orientations of the rectangular grid street

Fig. 1.4. Sketch of the Tower of the Winds in Athens.

pattern and the locations of entrances, courtyards and terraces associated with buildings in these cities. He also observed that some industries would be better sited with careful considerations to wind.

> 'Then let the directions of your streets and alleys be laid down on the lines of division between the quarters of two winds. On this principle of arrangement the disagreeable force of the winds will be shut out from dwellings and lines of houses. For if the streets run full in the face of the winds, their constant blasts rushing in from the open country, and then confined by narrow alleys, will sweep through them with great violence.'
> 'By shutting out the winds from our dwellings therefore we shall not only make the place healthful for people who are well, but also in the case of diseases due perhaps to unfavourable situations elsewhere, the patients who in other healthy places might be cured by a different form of treatment, will here be more quickly cured by the mildness that comes from the shutting out of the winds.'

Vitruvius' principles were taken up in the fifteenth century by Leon Battista Alberti, an Italian, who was influential in spreading these ideas from Italy to France, Germany, Spain and the rest of Europe including Russia [1.6].

In 1573 town planning laws were established for Spanish towns in South and Central America. New inland towns, according to these laws, were to be located on the eastern or western slopes of hills or mountains to shelter them from undesirable winds. Another provision urged that inland towns be exposed only to the north or south winds.

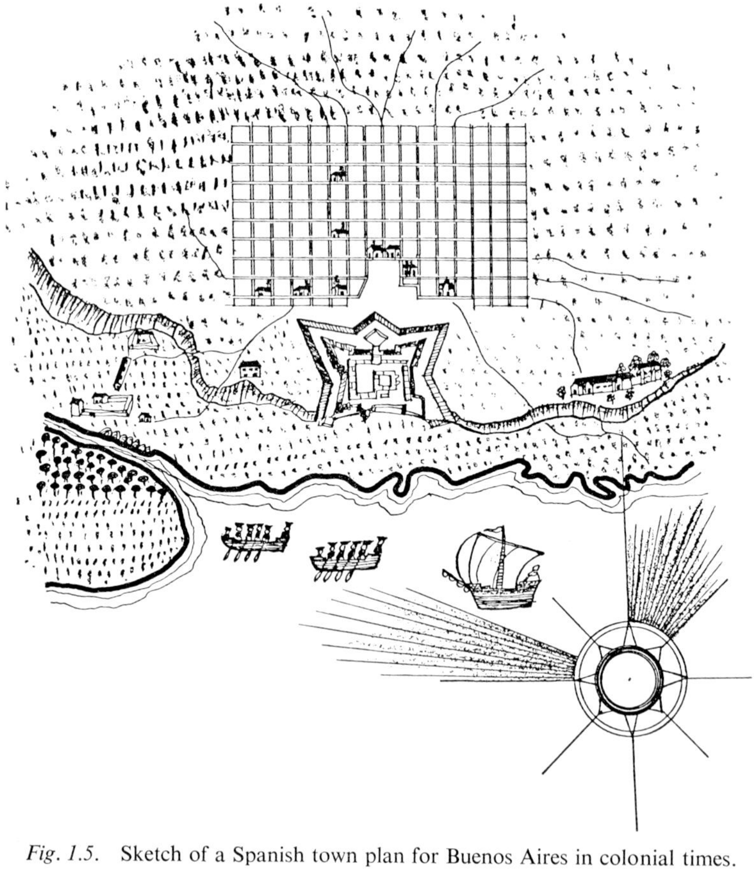

Fig. 1.5. Sketch of a Spanish town plan for Buenos Aires in colonial times.

Coastal towns were to avoid open water south or west of the settlement. This was a practical regulation because the prevailing winds along the Gulf Coast are south-easterly and would hinder the approach of sailing vessels to ports with channels leading into the wind [1.7].

The city of Buenos Aires was carefully planned around 1745 with the streets orientated to prevent the prevailing winds from sweeping through the length of the city (Fig. 1.5) [1.7].

Other cultures designed their buildings and towns to suit the local winds. The direction of the prevailing winds or storms governed the location of openings and entries in Japanese houses. It is still customary in Japan to consult a specialist on local climate when one wishes to build a traditional house [1.8]. Small openings on the northern side allow a minimal air current to pass through from north to south during colder months. Traditional houses are normally built on a slope, where possible, to make maximum use of slope winds for summer sliding screens although problems do occur in winter when cold winds penetrate the many cracks and crevices between these sliding door panels. This gives warmth in winter and coolness in summer (Fig. 1.6).

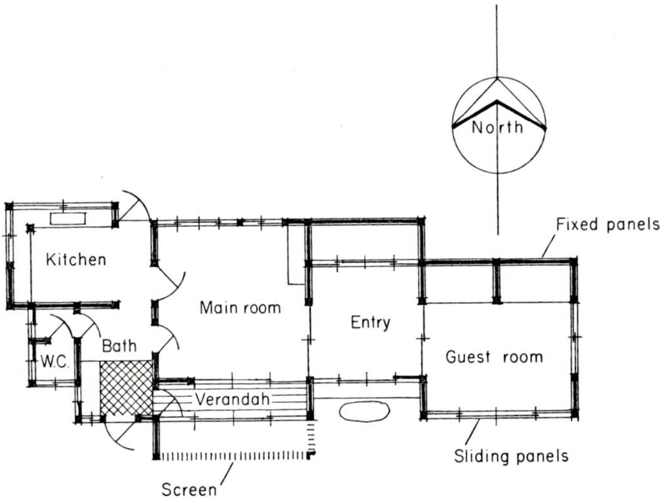

Fig. 1.6. Plan of a Japanese house indicating wall openings.

Early Canadian settlements took serious account of the winter conditions there. Houses near high ridges were sited on the southern side for protection from cold northerly winds. More exposed houses on the prairies had doors on the south side to avoid the strong north wind. Where front doors facing roads were orientated to the north they were kept permanently closed during the winter and entry was made through the kitchen doors at the rear. To minimise the entry

Fig. 1.7. Air scoops in Hyderabad.

of cold winds and loss of heated air, storm doors or air lock entries were provided at entrances to houses [1.9].

Hyderabad, a city in India, is characterised by numerous tall air shafts and wind scoops projecting from the roofs of buildings (Fig. 1.7).

Fig. 1.8. Sketch of the 'Mulguf' on an Egyptian house.

These rectangular chimneys have large flat panels over them which are inclined into the prevailing wind to deflect the roof top breezes down into the buildings. The induced airflow escapes through small windows into the narrow streets and alleys. A similar technique is evident in traditional Egyptian houses which use the 'Mulguf', a pair of roof ventilators in the form of triangular prisms (Fig. 1.8). The windward ventilator deflects breezes down into the house while the leeward ventilator acts as an exhaust [1.10].

1.2.1. Impact of the Industrial Revolution on Town Planning

The industrial revolution in Europe during the eighteenth century brought with it unprecedented urban growth and the need for city by-laws governing the natural ventilation and lighting of dwellings in rapidly overcrowding cities.

In the 1870s Nuremburg introduced laws regarding provision for natural light to each room in a house. A Swedish act passed in

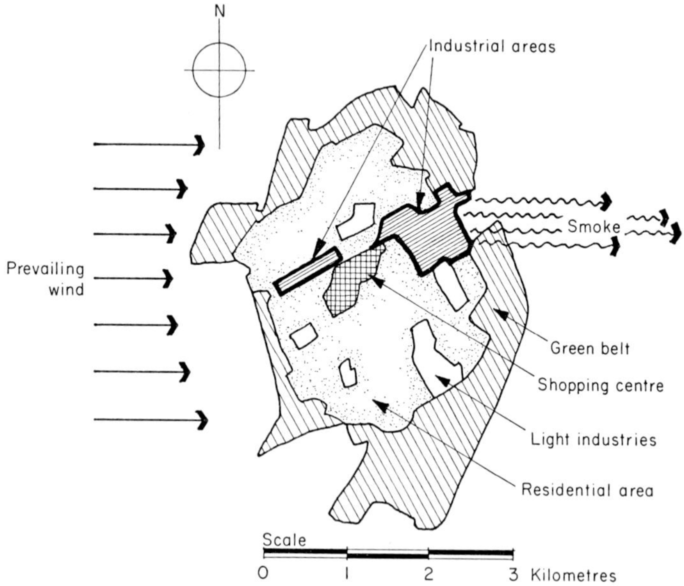

Fig. 1.9. Plan of the town of Letchworth indicating location of industrial areas and prevailing winds.

1874 required that light and air sufficient to maintain health in buildings be provided for in new town plans. Town planners in Vienna, around 1900, considered prevailing winds in the zoning of districts, the object being to ensure smoke from factories was blown away from the city instead of towards it [1.11]. Similarly the English garden cities of Letchworth and Welwyn have the industrial quarter on the east side of the town so the prevailing wind blows the smoke from the factories away from the town (Fig. 1.9). Russian town planner Muljutin, took similar precautions when he planned the city of Magnetogorsk [1.12].

More recent examples of the application of similar principles to town planning include Chandigarh, India and suburban planning,

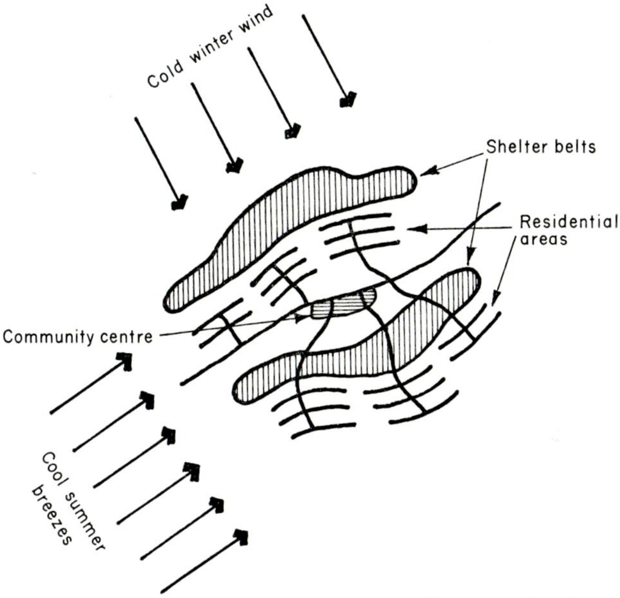

Fig. 1.10. Use of shelter belts in New Jersey towns for protection from winter winds.

New Jersey, U.S.A. Chandigarh's hot humid climate received careful attention by deliberate orientation of streets and spacing of buildings to ensure maximum airflow through housing to offset heat stress [1.13]. Olgyay [1.10] described the influence of wind effects in the planning of dormitory towns in New Jersey for people who commute to work in New York city. Large shelter belts of trees were cleverly planned to give protection from cold winter winds without obstructing the free passage for summer breezes which blow from a different quarter (Fig. 1.10).

1.3. Developments in Instrumentation

Prior to the seventeenth century attempts to design for wind used only the most general characteristics of the wind and it was not until instruments such as thermometers, barometers and anemometers were invented that any detailed analysis could be made from reliable climatic observations.

1.3.1. Wind Direction Indicators

Wind direction is equally as important as wind speed measurement. Early Chinese and Egyptian civilisations observed the wind direction from wind vanes. The Greeks also used wind vanes, one being mounted on their Tower of the Winds (Fig. 1.4).

1.3.2. Atmospheric Pressure Measurement

The invention of the barometer was the most important advance towards a scientific understanding of the origin and cause of wind. Credited to Evangelista Torricelli around 1643, it followed closely the invention of the thermometer by Torricelli's master Galileo.

Torricelli died soon after inventing the barometer, ending barometric experiments until Blaise Pascal (1623–1662) took up the investigations (Fig. 1.11). He sent his brother-in-law Perier up the Puy-de-Dom in Auvergne to observe barometric pressure changes and climbed St. Jaques-de-la-Boucherie in Paris himself to make similar observations. Others began following this practice and soon clambering up towers and mountains to watch the barometer reading fall, became one of the more popular outdoor sports among the amateur scientists of France.

Barometer design became a very popular activity. Barometers were devised in the form of wheels, spirals, some with attachments to assist in reading the scale. Experiments were made with various liquids replacing mercury, including cognac, the brainchild of a French enthusiast. Nevertheless the observations made with these instruments excited respected scientists of the day. Barometric observations established the fact that atmospheric pressure varied from day to day and from place to place and these pressure variations were connected with changes in weather (Fig. 1.12) [1.3].

1.3.3. Wind Speed Measurement

Wind speed measuring instruments are called anemometers and may be direct or indirect. Early *direct* measurements were made using windborne feathers and the shadows of clouds [1.14]. More recent direct methods include observing the velocity of particles in the air stream interpreted through Doppler effects on ground waves and electromagnetic radiation from radar and laser beams (Fig. 1.13) [1.15 and 1.16]. Most anemometers are of the *indirect* type. Leonardo

Fig. 1.11. Sketch of an experiment by Pascal and Petit with a water barometer at Rouen, in 1646.

da Vinci designed one but little development took place in wind speed measurement until Robert Hooke built his pendulum anemometer, similar to Leonardo da Vinci's, in 1667 (Fig. 1.14.).

Important advances in anemometer design were made by Pitot with the pressure-tube (Fig. 1.15), Bouguer and Osler with a spring and pressure plate anemometer (Fig. 1.16) and Lentmann with a device similar to a water wheel with buckets near the top of the wheel exposed to the wind. In 1846 Robinson used a vertical axle to which

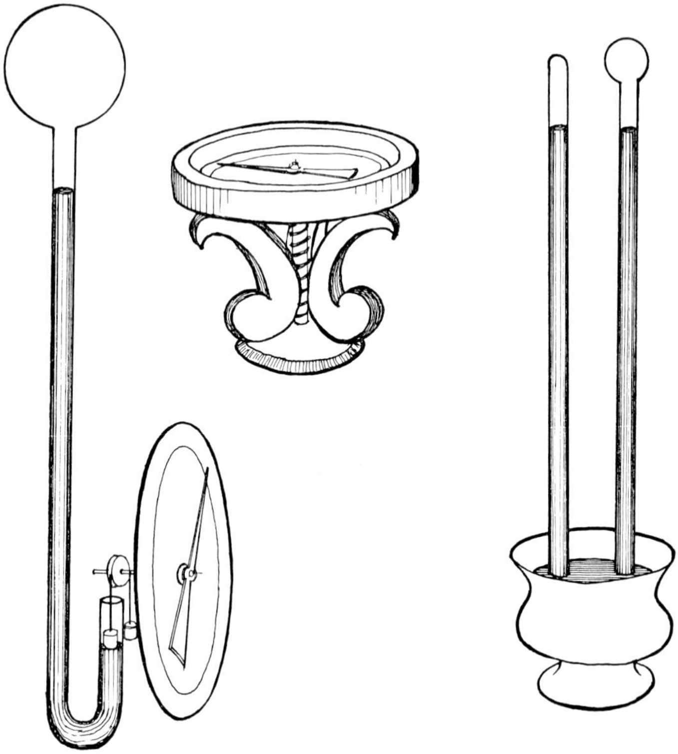

Fig. 1.12. Early barometers.

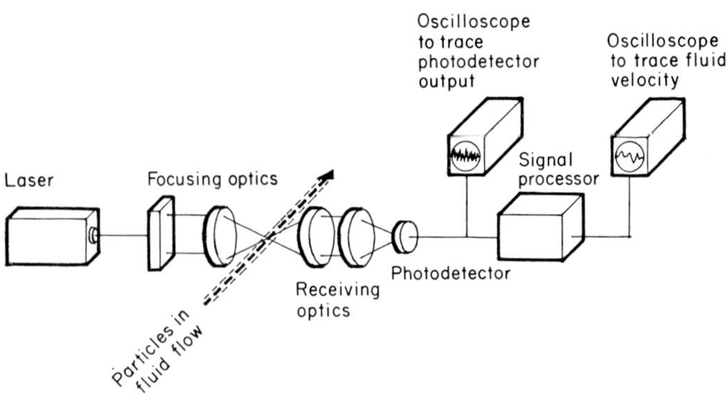

Fig. 1.13. Principal components of a laser Doppler anemometer.

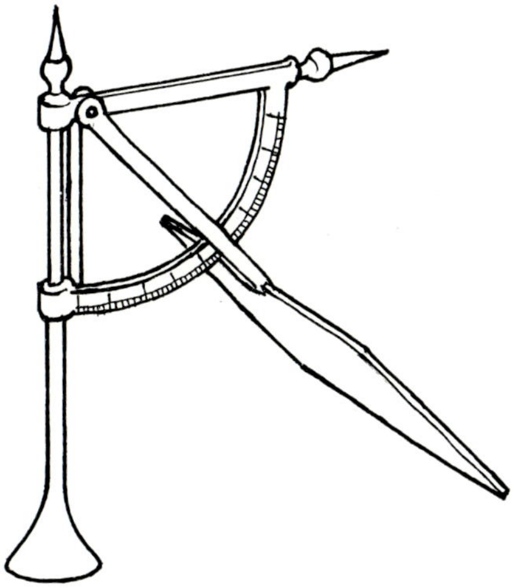

Fig. 1.14. Hooke's swing-plate anemometer.

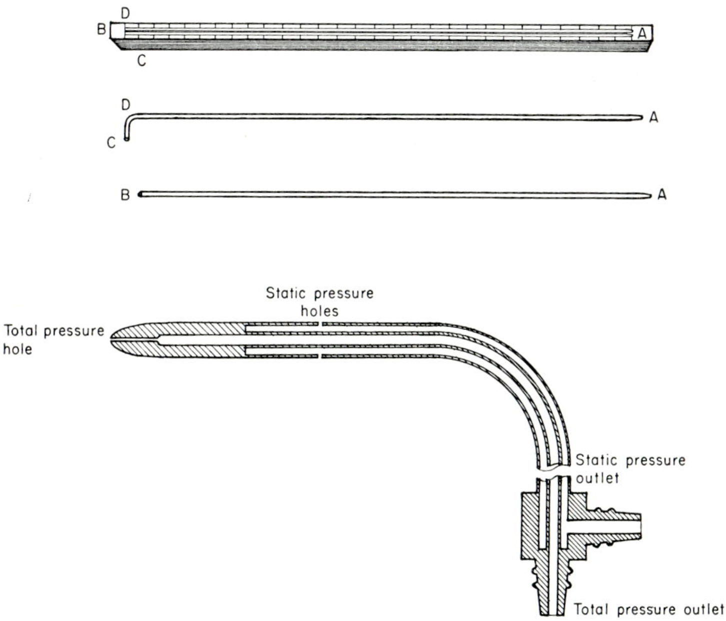

Fig. 1.15. Sketch of the original Pitot tube (above) and a current NPL type (below).

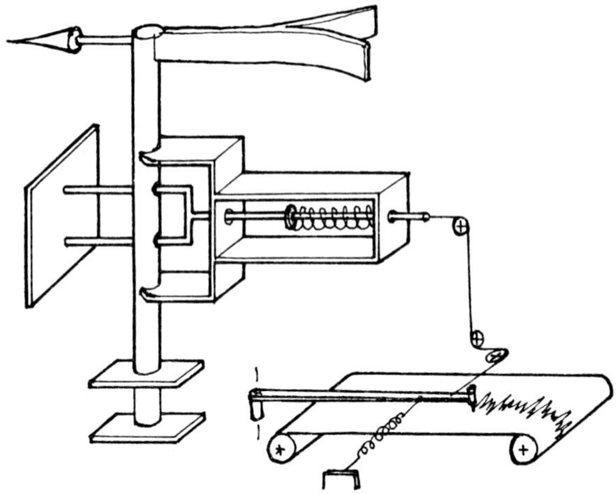

Fig. 1.16. Osler's pressure-plate anemograph.

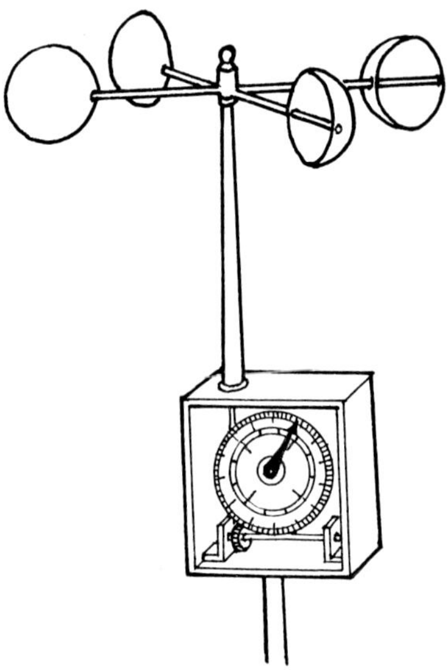

Fig. 1.17. Robinson's anemometer.

a number of cups were attached. These cups rotated in a horizontal plane; this was the prototype for the modern rotary cup anemometers (Fig. 1.17). Dines' anemometer of 1888–1889 incorporated Pitot's pressure-tube and a device for recording wind speed and direction. Recent developments in methods of indirect measurement include the use of cooling effects on heated wires and thermocouple junctions (Figs. 1.18 and 1.19).

Because different types of anemometers indicate different wind speeds, instruments must be selected with consideration to their response to various velocity ranges and gust sizes. International meteorological authorities originally standardised on the Dines pressure-tube anemometer (Fig. 1.20) but in recent years began introducing cup anemometers (Fig. 1.21). Both were mounted at a standard height of 10 m which enables direct comparison of wind data from numerous instruments around the world.

1.3.4. Wind Force Scales

In 1759 Rouse's table of wind pressures based mainly on experiments with windmills was read to the Royal Society by John Smeaton [1.17]. The wind speeds and wind forces were listed according to their 'common appellations' such as 'hardly perceptible' and 'very brisk' (Fig. 1.22). Admiral Beaufort developed a similar scale using nautical effects in 1806. This was later adopted by the Royal Navy [1.18]. A variation of this scale for land-based observers was later devised. This is still used as a guide to wind speed estimations but has been largely surpassed by wind speeds indicated by anemometers.

1.3.5. Wind Pressure Measurement

Like wind speeds, wind pressures change continuously due to turbulence and changing flow patterns around objects. Benjamin Baker (1840–1907), engineer in charge of construction of the Forth Bridge, measured local wind pressures on three panels, one with an area of approximately 28 m^2, the other two only 0·14 m^2 (Fig. 1.23). Higher wind pressures on the small panels were attributed to high energy gusts with dimensions closer to those of the smaller panels [1.19]. Introduction of this size factor into wind pressures on plates added further confusion to the problem of calculating wind forces.

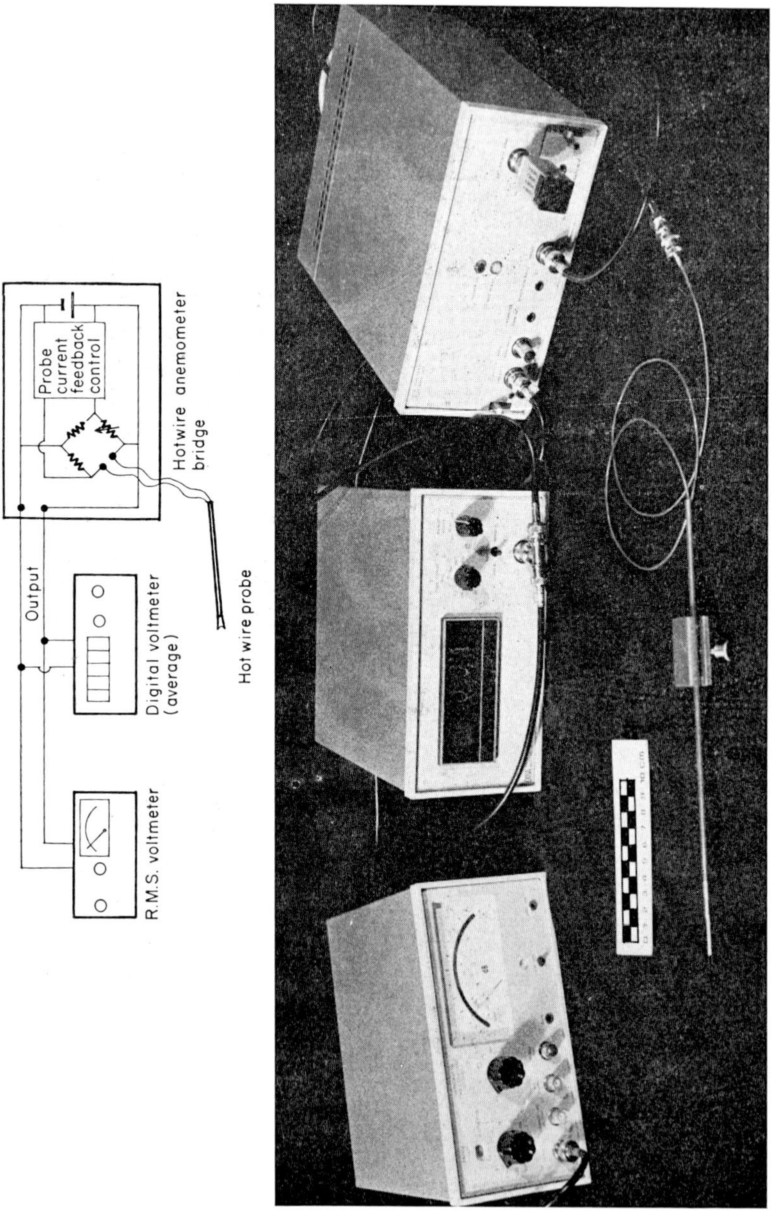

Fig. 1.18. Principal components of a hot wire anemometer system.

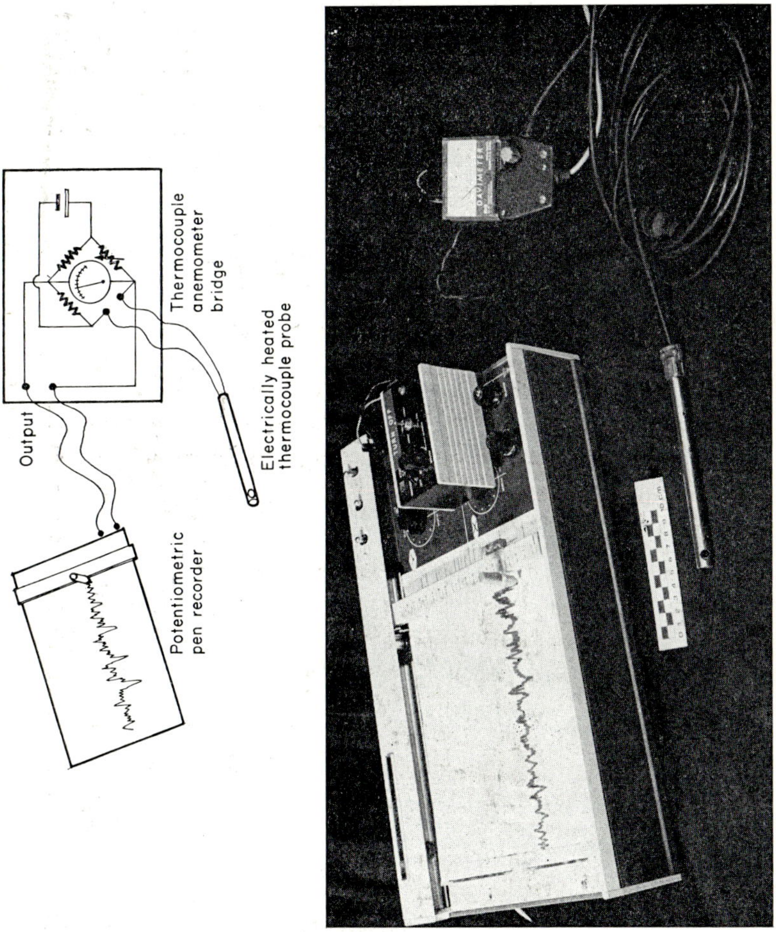

Fig. 1.19. A thermocouple type anemometer with a plotter indicating output signal.

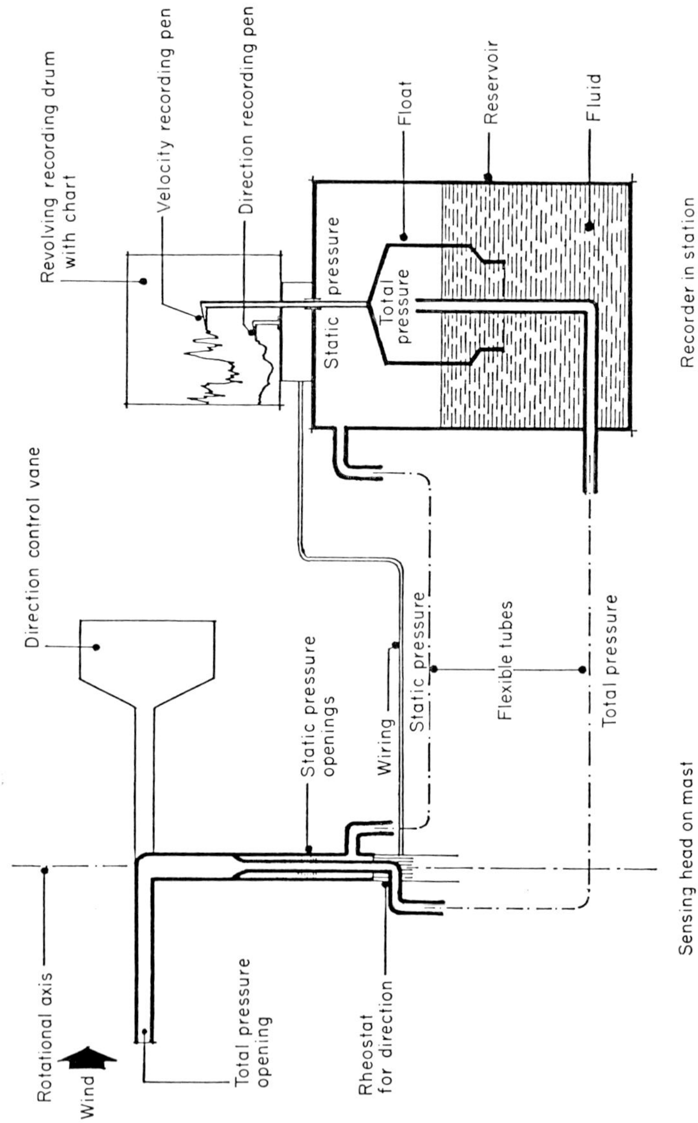

Fig. 1.20. Principal components of Dines pressure tube anemometer.

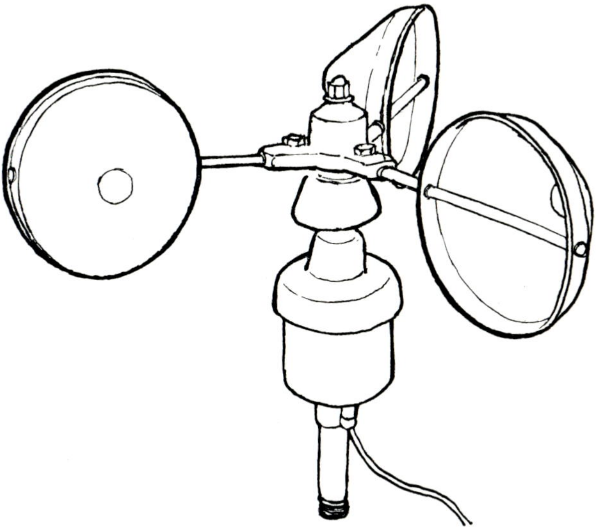

Fig. 1.21. A modern cup anemometer.

Velocity of the wind		Perpendicular force on one foot area (pounds avoirdupois)	Common appellations of the force of winds
miles/h	ft/s		
1	1·47	0·005	Hardly perceptible
2	2·93	0·020	Just perceptible
3	4·40	0·044	
4	5·87	0·079	Gentle pleasant wind
5	7·33	0·123	
10	14·67	0·492	Pleasant brisk gale
15	22·00	1·107	
20	29·34	1·968	Very brisk
25	36·67	3·075	
30	44·01	4·429	High winds
35	51·34	6·027	
40	58·68	7·873	Very high
45	66·01	9·963	
50	73·35	12·300	A storm or tempest
60	88·02	17·715	A great storm
80	117·36	31·490	A hurricane
100	146·70	49·200	A hurricane that tears up trees, carries buildings before it, etc.

Fig. 1.22. The velocity and force of wind according to their common appellations (after Smeaton, 1759).

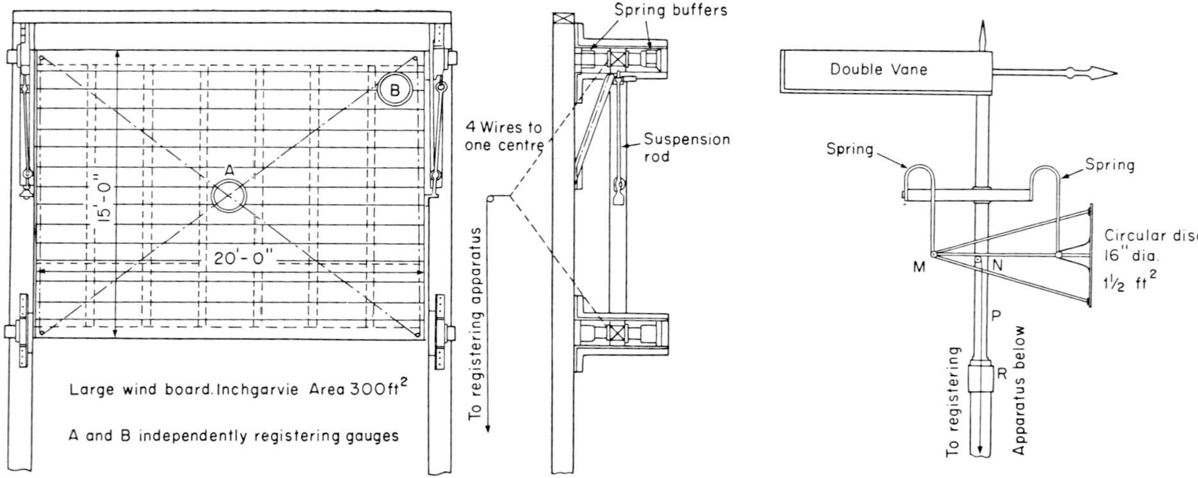

Fig. 1.23. Benjamin Baker's wind gauges used on Inchgarvie near the Forth Bridge.

A simple falling model method for comparing wind pressures on cylindrical structural sections with wind pressures on sections with flat surfaces was developed by Baker. Cardboard models of each section were fixed to opposite ends of a rod. After careful balancing, the rod was held horizontally and dropped a short distance. The end

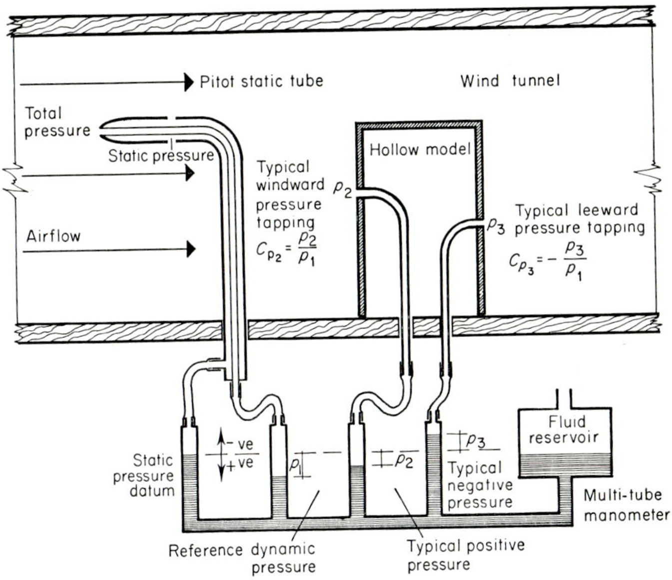

Fig. 1.24. Use of a multitube manometer to measure dynamic pressure of wind and surface pressures on a model in a wind tunnel.

of the rod to reach the ground first was subjected to less wind pressure or resistance. By varying the area of the flat surface section model an area of equivalent air resistance to tubular sections could be determined [1.19].

Pressure measurements using deflections of various pendulums proved of little value in determining wind pressures on structures and the practice of measuring pressures at numerous points on surfaces of models or full scale structures using U-tube manometers was established (Fig. 1.24).

Little change occurred until the development of electronic pressure transducers utilising strain gauges or capacitance and inductance

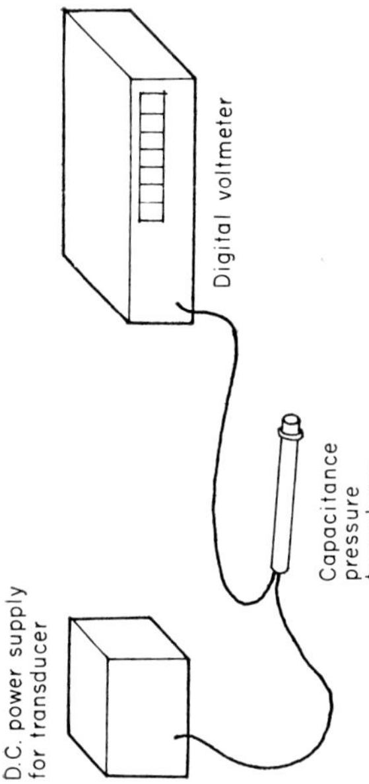

Digital voltmeter
Capacitance pressure transducer
D.C. power supply for transducer

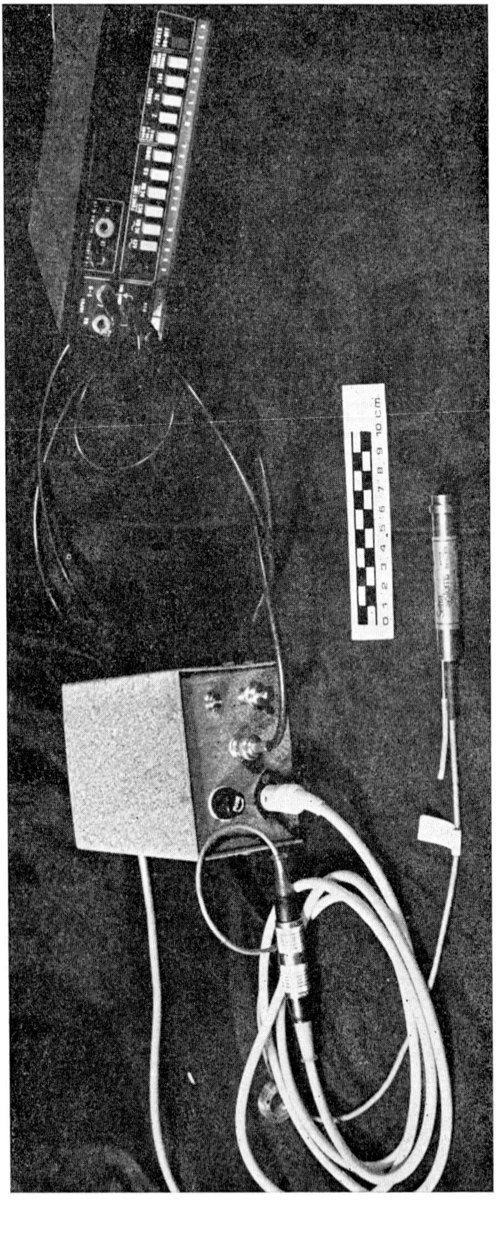

Fig. 1.25

effects (Fig. 1.25). These instruments allowed continuous monitoring and analysis of complex pressure by computers. Detailed specifications and characteristics of most current types of pressure and velocity measuring instruments are provided in a text by Ower and Pankhurst [1.20].

1.4. Developments in Fluid Mechanics

With the development of instruments to measure wind speeds and pressures, efforts were made to develop general theories to explain the behaviour of fluid motion.

Leonardo da Vinci's contribution to concepts of fluid flow was a result of his gifts for experiment and observation. His thoughts on flight and concepts of fluid mechanics are contained in his collected writings 'Del moto e misura dell'acqua' which includes sections on flow in pipes and eddies (Fig. 1.26). His techniques of flow visualisation by introducing opaque particles into the flow foreshadowed

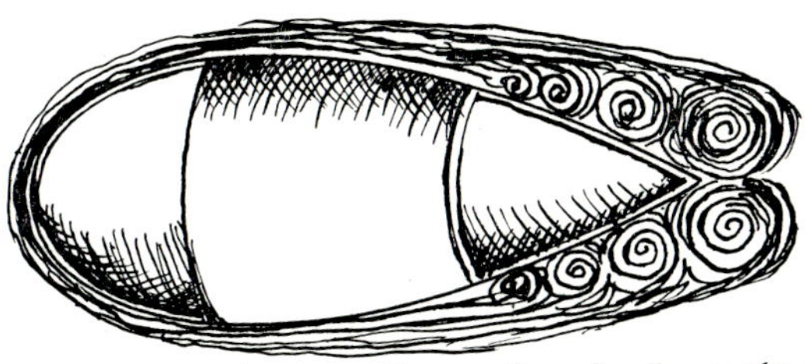

Fig. 1.26. Eddy formation in zones of separation after Leonardo da Vinci.

many of the classic experiments to be performed later in the nineteenth and twentieth centuries (Fig. 1.26) [1.21].

Fluid mechanics owes much to the work of Galileo and Newton in the seventeenth and early eighteenth centuries. Galileo proposed the law of uniform acceleration for falling bodies, as well as valuable contributions to instrumentation.

Newton provided the whole foundations of modern mechanics. His three proportionalities, between the force on a body in a fluid and its area, the square of its speed, and the fluid density were perfectly valid as far as they went. Since his time numerous investigators have concerned themselves with proving or disproving his formulae by experimental processes.

Newton also pronounced his law of viscosity by relating shear force or stress, viscosity, and relative motion of one layer of fluid adjacent to another layer. About the same point in time Blaise Pascal (1623–1662) put forward his axiom that, for fluids at rest, pressure is transmitted equally in all directions, now known as Pascal's law.

The foundations of classical hydromechanics began with Daniel Bernoulli (1700–1782) and Leonhard Euler (1707–1783). Bernoulli wrote his treatise 'Hydrodynamica', the first on the subject. This treatise included Bernoulli's principle, a fundamental theorem of fluid dynamics. 'In the case of steady flow of a fluid that is incompressible and non-viscous, the energy of a unit volume of a fluid is conserved along a streamline. More specifically to be expressed as the algebraic sum of the three terms, the pressure, kinetic energy per unit volume, potential energy per unit volume, has a same value at any two points on a streamline.'

Euler developed the general non-viscous equations of motion in terms of the acceleration components of a fluid particle. By applying Newton's second law, the equation of motion, to a fluid particle in the direction along the streamline in steady incompressible frictionless flow, Euler produced a differential equation which now bears his name [1.22].

The nineteenth century saw many contributions to the understanding of fluid flow. The Italian, Venturi, studied the flow of fluids through conical reductions and expansions in tubes for the purpose of reducing the turbulence and losses caused by changes in pipe cross-sections.

A Frenchman, Navier, extended the Euler equation to include the attraction of adjacent molecules. Later an Englishman, Stokes, made a final derivation of the equations of motion in terms of the coefficient of viscosity. These are now known as the Navier–Stokes equations [1.23]. Solutions to these equations are at present limited to a few simple flow geometries which severely restricts their application in architectural aerodynamics.

Barré de Saint-Venant developed equations of motion of a fluid particle in terms of the shear and normal forces exerted on it.

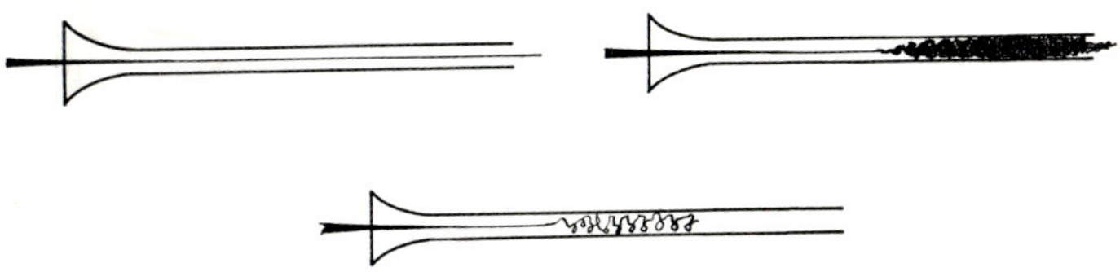

Fig. 1.27. Sketches, after Reynolds, of the transition from laminar to turbulent flow in pipes.

Poiseuille determined the equation for laminar flow through capillary tubes, while von Helmholtz and Lord Kelvin made important contributions to the knowledge of vortex motion.

Perhaps the most important contribution to the reconciliation of theoretical and observed fluid behaviour was made by the British physicist, Osborne Reynolds (1842–1912) who made important advances with the development of the dimensionless parameter now called Reynolds number. This number was used to distinguish laminar from tubulent flow in a closed conduit [1.24] (Fig. 1.27). He also expressed the Navier–Stokes equations in a form introducing the concepts of a turbulent shear.

The twentieth century has seen contributions from numerous investigators, as the field of fluid research broadened. Some contributors stand out such as Ludwig Prandtl (1875–1953) who in 1904 stated that for fluids of low viscosity, the effects of viscosity may be considered to be limited to a thin layer at the boundaries of the flow [1.25]. Outside this boundary layer irrotational flow exists. Prandtl's experiments and insights into the mechanics of fluid flow together with those of his student, von Karman, developed the knowledge of boundary layer velocity distributions and the phenomena of vortex shedding in wakes (Fig. 1.28), two major contributions to the science of architectural aerodynamics [1.26].

Advances were also being made in the application of fluid mechanics to the estimation of natural ventilation. During a fellowship at Cambridge College in 1903, Napier Shaw gave a series of lectures in which he described the behaviour of air currents and laws

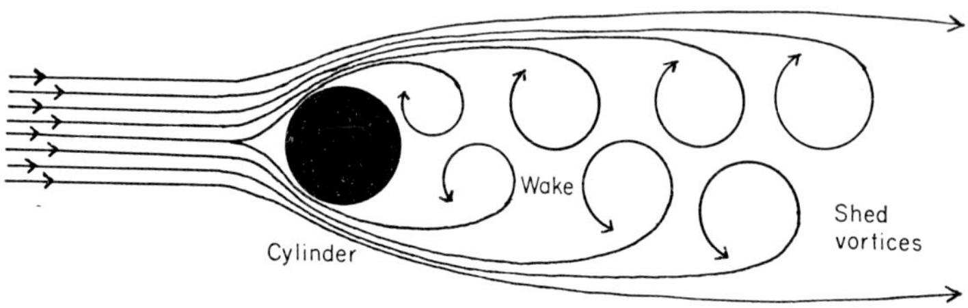

Fig. 1.28. Flow pattern around a circular cylinder indicating vortex shedding downstream in the wake.

of ventilation using electrical circuit analogies. The equation suggested by Shaw required knowledge of the total pressure across the orifice and a discharge coefficient or efficiency of the opening [1.27].

A growing interest in infiltration due to variations in the density of heated air in tall buildings in the United States, resulted in the American Society of Heating and Ventilating Engineers (A.S.H.V.E.) sponsoring a series of research projects for which reports were published between 1924 and 1933 [1.28]. These reports provided infiltration rates for various types of windows per unit pressure difference. Later papers published by A.S.H.V.E. members described results of experiments on infiltration due to air density variations in tall buildings [1.29].

By 1950 the Building Research Station in England had taken an active interest in infiltration in housing. An account of these studies on an experimental house at the station was given by Dick [1.30] in which measurements were found to be in good agreement with theoretical estimates using pressure differences from wind tunnel studies

1.5. Developments in Meteorology

The science of meteorology advanced rapidly after the invention of the barometer. Torricelli had observed that on warm spring days in Italy a cold stream of air flowed out through the wide doors of large churches. He assumed that the air in the large buildings was, at that time, appreciably cooler and heavier than the neighbourhood and therefore flowed out of doors.

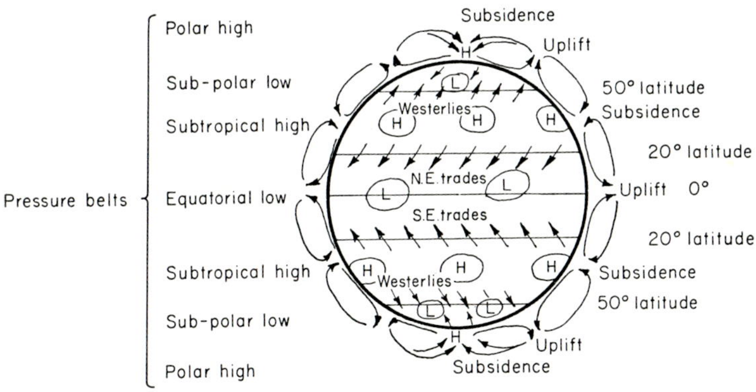

Fig. 1.29. Characteristics of global atmospheric circulation.

Better understanding of weather conditions resulted from Ferdinand II's establishment of an International Meteorological Organisation in the seventeenth century. Temperature and barometric pressure readings were taken, usually by Jesuit priests in various European cities. These were compared. This organisation continued after Ferdinand's death and eventually led to Halley's discoveries in the later seventeenth century that: 'According to the law of staticks, the air which is less rarefied and expanded by heat and consequently more ponderous must have motion towards those parts thereof which are more rarefied and less ponderous to bring it to equilibrium'. This led Halley to correctly explain the Monsoon and Trade Winds [1.3].

Halley, however, had not been able to explain why the Trade Winds blew from the northeast in the northern hemisphere and from the southeast in the southern hemisphere. George Hadley in the early eighteenth century explained these deflections as being due to the rotation of the earth (Fig. 1.29) [1.31]. Benjamin Franklin acknowledged Hadley's theories in 1753 although it took nearly 100 years before it was widely accepted by the meteorologists.

Franklin was the first to discover that storms move with their own inner circle of winds and do not move in the direction of the wind they bring with them. This discovery was a landmark in meteorology.

Meteorological theory has been greatly assisted by the introduction of new instruments, and the standardisation of their application. Twentieth century developments in automatic radio relaying of weather information from remote weather stations, together with world weather satellite coverage are providing unprecedented opportunities for understanding and predicting world weather including winds.

1.6. Simulated Wind Studies

The variable nature of natural wind forced early experimenters to create artificial airflow around their models. In 1746 Robins read two papers to the Royal Society under the titles 'Resistance of the Air' and 'Experiments Relating to Air Resistance'. These papers described his invention, a whirling table or arm apparatus used to measure the wind forces on flat surfaces and geometric bodies which he used to improve the shape of projectiles for military purposes (Fig. 1.30).

John Smeaton (1742–1792) civil engineer and his friend Rouse concerned themselves with wind speeds and pressures, providing valuable data for builders, bridge designers, architects, ship designers and for a brief period those interested in aeronautical matters. Smeaton improved Robins' whirling arm and with it tested models of windmill sails (Fig. 1.31).

Engineering literature of the period was inundated by numerous conflicting empirical equations for determining wind pressures from wind speeds. Much of this research was prompted by wind failures such as Tay Bridge [1.32] and concern for other structures which exploited the tensile strength of steel but apparently lacked sufficient wind bracing and the mass of earlier masonry structures.

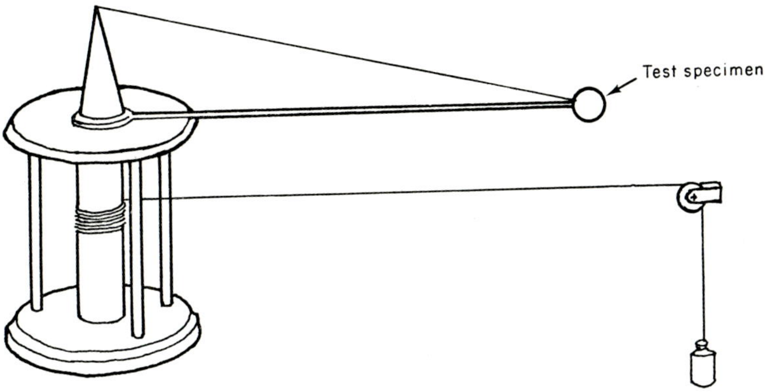

Fig. 1.30. Sketch of the whirling arm used by Robins.

The first wind tunnel appears to be one constructed by Francis Wenham when commissioned by the Aeronautical Society of Great Britain to study wind pressures on flat plates inclined to the wind. A description of the apparatus and results obtained were published in the Society's Annual Report in 1871. A fan blower was used to deliver air through a timber duct approximately 3 m long with a square cross-section with sides approximately 0·5 m. Lift and drag forces on inclined plates were measured by spring deflections off the

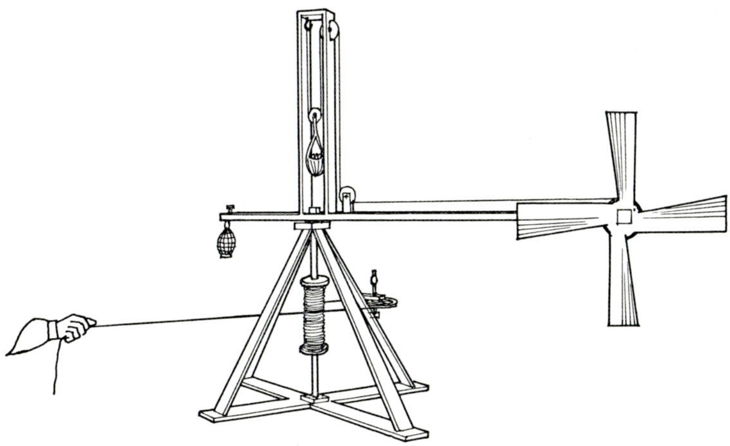

Fig. 1.31. Smeaton's adaptation of Robins' whirling arm for windmill tests.

mounting arm supporting the plates. Results of Wenham's experiments on inclined plates were used by Professor Unwin to calculate wind pressures on roofs. An influential structural engineering text of the time by Stoney refers to Unwin's design wind pressures for roofs [1.33].

1.6.1. Early Wind Tunnel Studies of Model Buildings

The earliest wind tunnel investigations of wind pressure on model buildings appear to be those carried out by Professor W. C. Kernot in Australia and J. O. V. Irminger in Denmark. In the early 1890s Professor Kernot, at Melbourne University, constructed a small open jet wind tunnel and studied wind forces on a wide range of geometrical forms, including model buildings. Results of these studies were published in October 1893 [1.34]. Airflow for the wind tunnel was provided by a propeller 0·7 m in diameter with a pitch of 1·2 m, driven by a gas engine at 400–800 rev/min. Airflow from the propeller passed through a short circular duct fitted with a radial diaphragm into a funnel with an outlet cross-section approximately 0·3 by 0·25 m, to produce a steady jet of air across a test table. Models were mounted on a vertical stem attached to a delicate carriage which was free to move on a level surface of the test table. The wind force on each model was measured by a spring balance attached to the carriage (Fig. 1.32). Building studies by Kernot included the effect of parapets, roof pitch and an absence of various walls on wind forces on roofs.

Irminger in 1893 performed wind pressure studies using a gas works' smoke stack into which a timber test section approximately 1·02 m in length and 0·23 by 0·11 m cross-section was installed. These studies included pressure distributions on flat plates, a primitive airfoil section, cylinders and simple building models (Fig. 1.33). Results of these studies were published in 1894 [1.35].

1.6.2. Introduction of Thick Boundary Layers in Wind Tunnels

Further studies by Irminger and Nokkentved in Denmark [1.36] and Bailey [1.37] in Great Britain in the 1930s indicated significant discrepancies between wind tunnel pressure distributions on models in uniform flow and full scale studies on similar buildings (Fig. 1.34). In a further experiment, results of which were published in 1943,

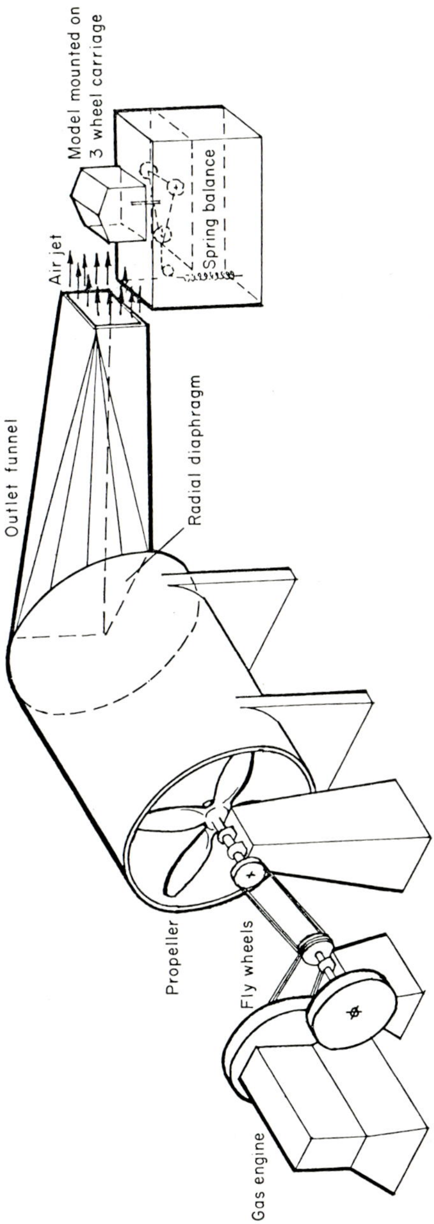

Fig. 1.32. Schematic layout of Kernot's wind tunnel from a description in his paper [1.34].

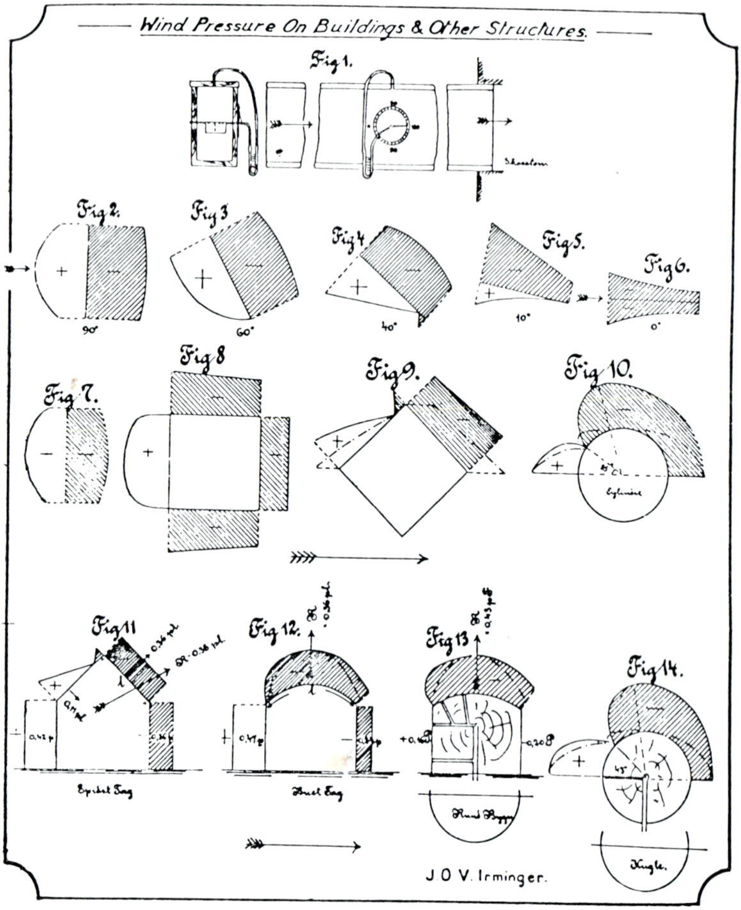

Fig. 1.33. Sketches of Irminger's apparatus and pressure distributions (published 1894) [1.35].

Bailey and Vincent obtained much better agreement between model pressures and full scale observations using a very long wind tunnel in which a thick boundary layer was generated [1.38]. Similar techniques were developed in 1954 by Jensen [1.39].

Applications of wind tunnel testing in the 1950s soon expanded from wind loading considerations to other aspects of wind effects such as wind shelter provided by buildings, trees, hedgerows and fences, by Jensen [1.39], in Denmark, Sato *et al.* [1.40] in Japan and White [1.41] at the Texas Engineering Experiment Station.

Wind tunnel studies of natural ventilation due to wind forces were pioneered at the Texas Engineering Experiment Station [1.42 and 1.43], in a programme to improve the design of school buildings. Van Straaten [1.44 and 1.45] conducted similar studies in South Africa and developed a method of estimating natural ventilation using wind pressure distributions on the building and coefficients for the efficiency of horizontally pivoted window openings.

It was not until 1958 that Jensen formulated his 'Model' law [1.46]. 'The correct model test for phenomena in the wind must be carried out in a turbulent boundary layer and the model-law requires that this boundary layer be to scale as regards to velocity profile.'

This modelling requirement eliminated the possibility of testing model buildings in aeronautical wind tunnels without drastic modifications to the flow, as the flow in these tunnels was extremely uniform with very low turbulence. Aerodynamicists eager to assist building designers had often tested buildings in these tunnels. Fortunately the data gained, although varying greatly from the real values, resulted in loads within the large safety factors existing in building design at that time. The rapid acceptance of Jensen's model-law quickly put an end to wind tunnel studies of buildings in uniform flow. From this point in the early 1960s new wind tunnels began to emerge with facilities to create appropriate increases in mean wind speed with height above the tunnel floor. These velocity profiles were induced using a number of techniques including curved screens, graded screens, vorticity generators and roughness elements on the tunnel floor.

Pressure distribution studies performed with carefully modelled velocity profiles resulted in considerable differences in pressure distributions from those previously observed in uniform flow. A significant difference in these distributions was the shift in high pressure region on the windward face towards the top of the building first noted by Bailey [1.37]. Further studies of these differences were made by Baines [1.47] and Hamilton [1.48] (Fig. 1.35).

By the early 1960s uncomfortable wind conditions for pedestrians around the base of tall buildings and in arcades were studied by

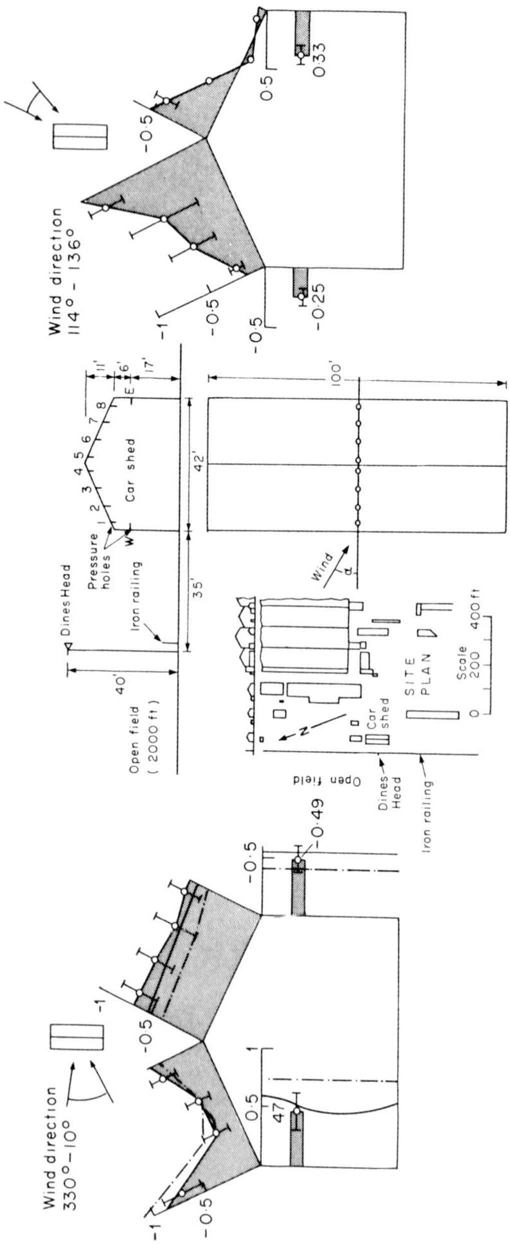

Fig. 1.34. Wind pressures on car shed by Bailey (1933).

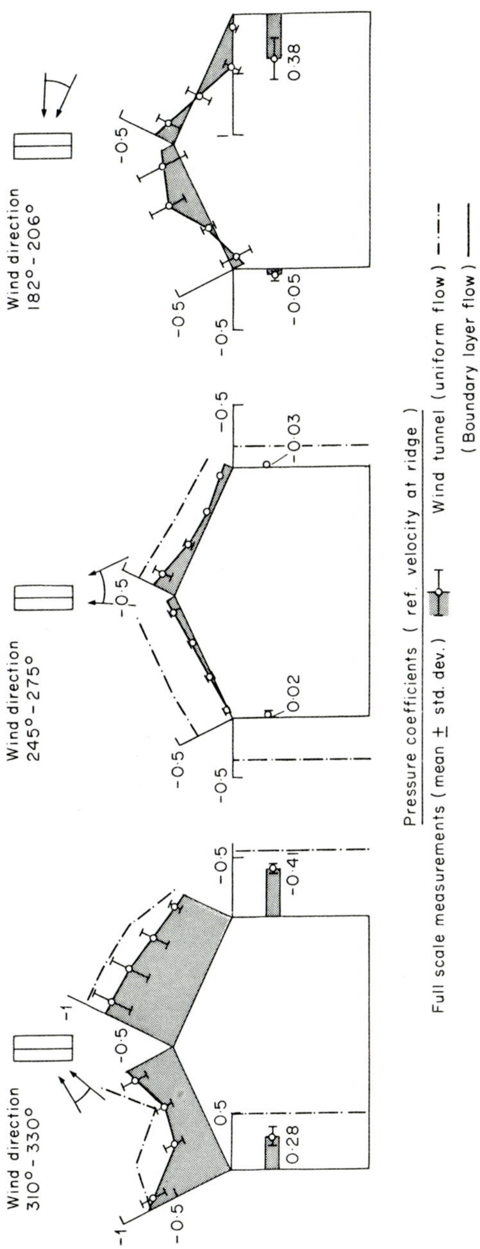

Fig. 1.34—contd.

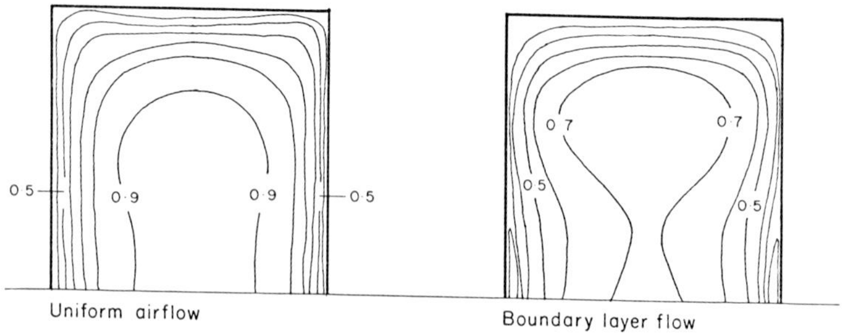

Fig. 1.35. Pressure distributions in pressure coefficient form on the windward face of a cube in uniform flow and typical boundary layer flow (after Hamilton).

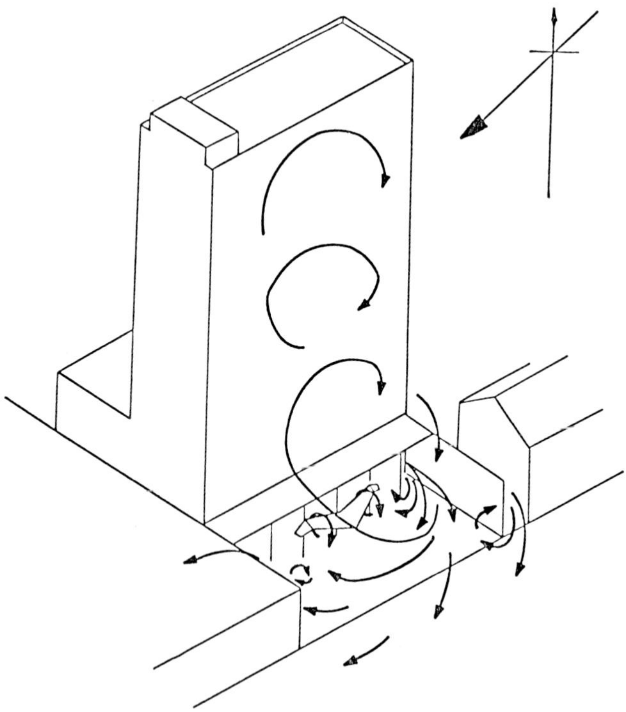

Fig. 1.36. A sketch of one of many three dimensional airflow patterns around a city building with a vehicular driveway through it (after Trimble).

Trimble [1.49] and Davenport [1.50], the Building Research Station in Great Britain [1.51] and Cermak [1.52], Brisker, O'Hare and others in the United States [1.53 and 1.54]. The earliest report by Trimble [1.49] clearly showed by isometric sketches the three dimensional airflow patterns near a tall building with a vehicular passageway through its base and the effects of a range of screens and hoods to protect pedestrians from local wind effects (Fig. 1.36).

1.7. Full Scale Studies of Wind Effects

Generally the difficulties in instrumentation and the expense associated with lengthy duration of full scale studies has limited their application to essential correlation studies to verify wind tunnel observations. These have tended to be carried out by government research organisations such as the Building Research Station (now Building Research Establishment) [1.55], the National Research Council, Canada [1.56] and the U.S. National Bureau of Standards [1.57]. Joubert performed the first full scale wind pressure distribution study in Australia on a building at the University of Melbourne [1.58].

Some problems, such as wind generated noise [1.59], wind effects on heat transfer through external walls, and infiltration in buildings through small cracks and openings [1.60] are difficult to simulate using models. Another important area, for full scale studies, is comfort surveys to develop wind criteria for the many human activities that take place in city and suburban spaces [1.61].

Development of design data in areas involving full scale studies has been slower than in other areas where models can be used. Interested universities and other teaching institutions have taken an interest in this type of research [1.62 and 1.63]. These organisations are providing valuable assistance to the government agencies working in this area.

1.8. Effect of Tall Building Development on Architectural Aerodynamics

It was during the 1950s that significant changes occurred in the design and construction of tall buildings. Refined structural analysis of steel frames together with lightweight flooring and external cladding materials produced lighter weight, more flexible buildings with less damping than earlier steel framed buildings which retained a large amount of masonry construction. This new breed of tall buildings was sensitive to dynamic excitation by wind, and unnerving building motions during strong winds forced building designers to seek solutions using the newly developed boundary layer wind tunnels.

Complex energy distributions of natural wind due to its turbulence presented a formidable problem to be overcome. Significant breakthroughs by Davenport and others were made in the 1960s in methods of rationalising the complex characteristics of natural wind together with wind tunnel determination of a building's response to wind gusts, to a point where existing dynamic structural analysis techniques could be used. A rapid growth in the number and height of tall buildings was accompanied by a similar increase in the number of researchers investigating wind-induced movements in buildings. Limiting accelerations to a level below the perception of occupants soon became a prime structural design criteria for tall buildings.

In 1963 the first symposium devoted solely to the effects of wind on buildings and structures was held at the National Physical Laboratory, Teddington, in Great Britain. The success of this symposium and the publishing of its 850 pages of proceedings [1.64] saw the establishment of an active group, the International Wind Study Group. This group has now been reformed under the new title The International Association for Wind Engineering. Further international conferences were organised in Canada during 1967 [1.65], Tokyo during 1971 [1.66] and London during 1975 [1.67]. These conferences accelerated developments in architectural aerodynamics by providing a forum for detailed discussion and appraisal of the techniques and theories being developed at that time.

Much of the discussion of wind effects conferences was related to structural considerations. However, by the 1971 conference in Tokyo serious consideration was given to a number of environmental factors. Problems of strong and gusty wind conditions in pedestrian spaces around tall buildings had emerged as a serious problem with the death of a woman in Great Britain. More recently problems related to high local suctions due to vortices close to the building surfaces have become the subject of long term detailed full scale studies as well as conditions needed to reproduce these effects in wind tunnel model studies [1.68].

1.9. Current Directions in Architectural Aerodynamics

It has become apparent in the last few years that much of the early data gathered from wind tunnel studies will need to be replaced with updated information to account for a range of velocity profiles and turbulence characteristics. Some wind effects can now be satisfactorily estimated using existing theory of fluid mechanics or empirical relationships. The complex and changeable characteristics of atmospheric conditions and the almost infinite variety in the geometry of built forms, produce an endless variety of wind effects. It will probably take many more years of carefully conducted wind tunnel studies together with full scale correlation to produce sufficient information to allow a significant increase in theoretical means of estimating wind effects. Until these are developed careful wind tunnel studies are the only practical means of obtaining much of the data needed in the field of architectural aerodynamics.

Chapter 2

Basic Aerodynamics

Before proceeding to the more specific chapters which discuss aerodynamic flow fields, pressures and loads, it is necessary to develop some understanding of basic aerodynamics. This does not mean that a complete knowledge of the fundamentals of fluid mechanics are required. In fact, for the study of architectural aerodynamics, it is more important to understand the behaviour of real fluids which are dominated by the effects of viscosity and turbulence and feel free to move in three dimensions. Hence, only enough of the fundamentals to permit an understanding of real fluid behaviour will be introduced in this text.

2.1. Flow Fields and Regions

Wind flow near the earth's surface is turbulent, that is gusty, and unsteady. By unsteady it is meant that variables such as velocity and pressure at a fixed point in space are varying with time. In the case of turbulent flow, this variation is random and the process is called a stochastic process. Fortunately, it has been found possible to separate the turbulent flow properties into mean and fluctuating parts. For example, the longitudinal velocity component, u, can be described as the sum of the mean, $\bar{u}$, and the deviation from the mean, u',

$$u = \bar{u} + u' \qquad (2.1)$$

For most of the phenomena which occur in architectural aerodynamics the mean part can be regarded as steady with the unsteady, time dependent, part being incorporated in the fluctuating part. Hence, much of the flow and interaction phenomena can be usefully described in terms of the steady, time averaged mean characteristics.

Flow fields can be conveniently divided into three flow regions, as shown by several examples in Fig. 2.1, and can be described as follows:

(1) Freestream flow
The freestream flow is the flow ahead of and outside of the immediate influence of the body. Most importantly, this is the region where mean shear rates are negligibly small, and where the Bernoulli equation, shortly to be introduced, is valid.

(2) Shear layers
Shear layers are regions of relatively high mean shear rate. A boundary layer of fluid through which the mean velocity changes from zero at a solid boundary to free stream velocity at the outer edge is a very important shear layer. When a boundary layer separates from a surface or at an edge, it becomes a free shear layer separating freestream flow on one side from a wake or re-circulating flow region on the other side.

(3) Wake flow
The wake region developed behind a separating shear layer contains low velocity (relative to freestream) re-circulating eddies, or vortices as they are sometimes called. Two boundary layers which meet at the trailing edge of a streamlined shape also develop a type of wake flow.

2.2. Streamlines and Continuity

A streamline is defined as a line which is everywhere tangential to the instantaneous direction of flow. For steady flow conditions the path of a trail of smoke is coincident with the streamline. Hence, it is possible to use smoke or dye tracers to provide visualisation of a steady streamline pattern, of which the streamlines used to illustrate the freestream flow regions in Fig. 2.1 are typical of what can be observed. A similar observation of the flow can be made using wool tufts. A time lapsed photograph of wool tufts can also be used to illustrate the magnitude of the turbulence, or unsteady component. In Fig. 2.2 a grid of wool tufts has been arranged to show the mean streamlines in the flow around a flat circular disk, the degrees of unsteadiness in and near the wake and the reversed flow immediately behind the disk.

One of the most important features about streamlines is that they

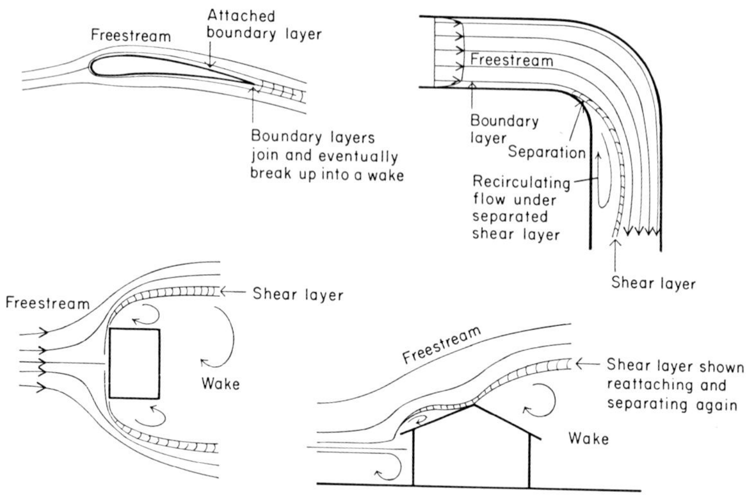

Fig. 2.1. Examples of mean flow fields with attached and separated flows.

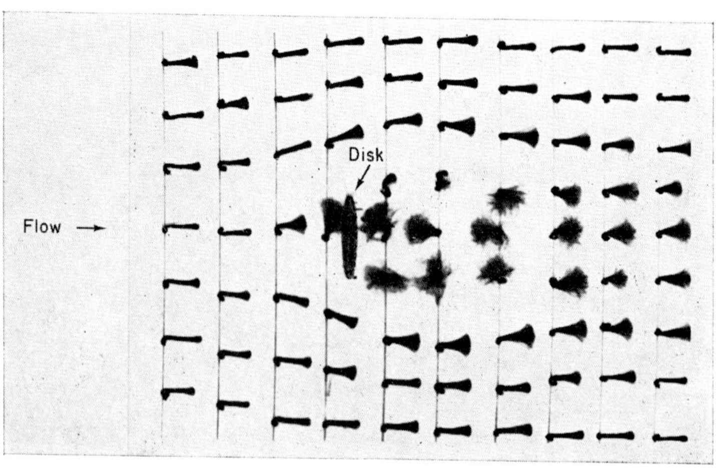

Fig. 2.2. Flow around a flat circular disk illustrated by a grid of wool tufts.

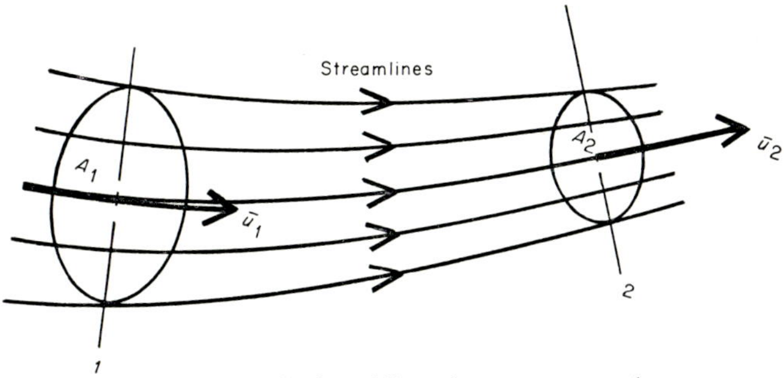

Fig. 2.3. Continuity of flow along a stream tube.

can be used to define the boundary of a stream tube. By definition there can be no flow across the bounding streamlines and hence a simple statement of the principle of continuity in relation to two planes normal to the stream tube in Fig. 2.3 can be made as follows:

$$\text{mean mass flow rate} = \rho_1 A_1 \bar{u}_1 = \rho_2 A_2 \bar{u}_2 \tag{2.2}$$

and for incompressible flow.

$$\text{mean volume flow rate, } Q = A_1 \bar{u}_1 = A_2 \bar{u}_2 \tag{2.3}$$

where ρ is the fluid density, A is the area of the normal plane in the stream tube, $\bar{u}$ is the mean velocity across the plane.

This principle of continuity gives a convenient mean velocity relationship along a stream tube or bounded flow whose area is varying,

$$\frac{\bar{u}_1}{\bar{u}_2} = \frac{A_2}{A_1} \tag{2.4}$$

2.3. Bernoulli's Equation and Pressure Fields

The relationship between pressure and velocity at different parts of a fluid field can be obtained by an analysis of the dynamics of a particle of the fluid. The complete analysis is very complex but for freestream flow a number of simplifying assumptions can be made to produce what is known as Bernoulli's equation which, in terms of developing an understanding of fluid behaviour, is probably the most important expression in fluid mechanics.

The assumptions made are that the flow is,

(1) steady (no time dependent variables);
(2) inviscid (viscous forces are negligible);
(3) irrotational (elements of the fluid undergo zero rotation as they deform in the flow);

and then Bernoulli's equation with reference to Fig. 2.4 may be written

$$p_1+\tfrac{1}{2}\rho\bar{u}_1{}^2+\rho g z_1 = p_2+\tfrac{1}{2}\rho\bar{u}_2{}^2+\rho g z_2 \tag{2.5}$$

where p is the static pressure, $\bar{u}$ is the mean velocity, z is the height above a horizontal reference datum, g is the gravitational acceleration, ρ is the fluid density.

For subsonic aerodynamic problems generally the gravitational term is usually negligibly small, the air is regarded as incompressible and the density constant (except for the effect of temperature). This means that an increase in velocity is accompanied by a drop in pressure and vice versa, as given by

$$p+\tfrac{1}{2}\rho\bar{u}^2 = \text{constant} \tag{2.6}$$

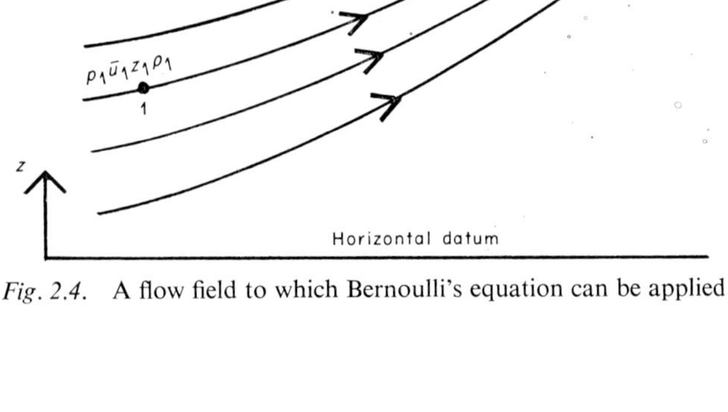

Fig. 2.4. A flow field to which Bernoulli's equation can be applied.

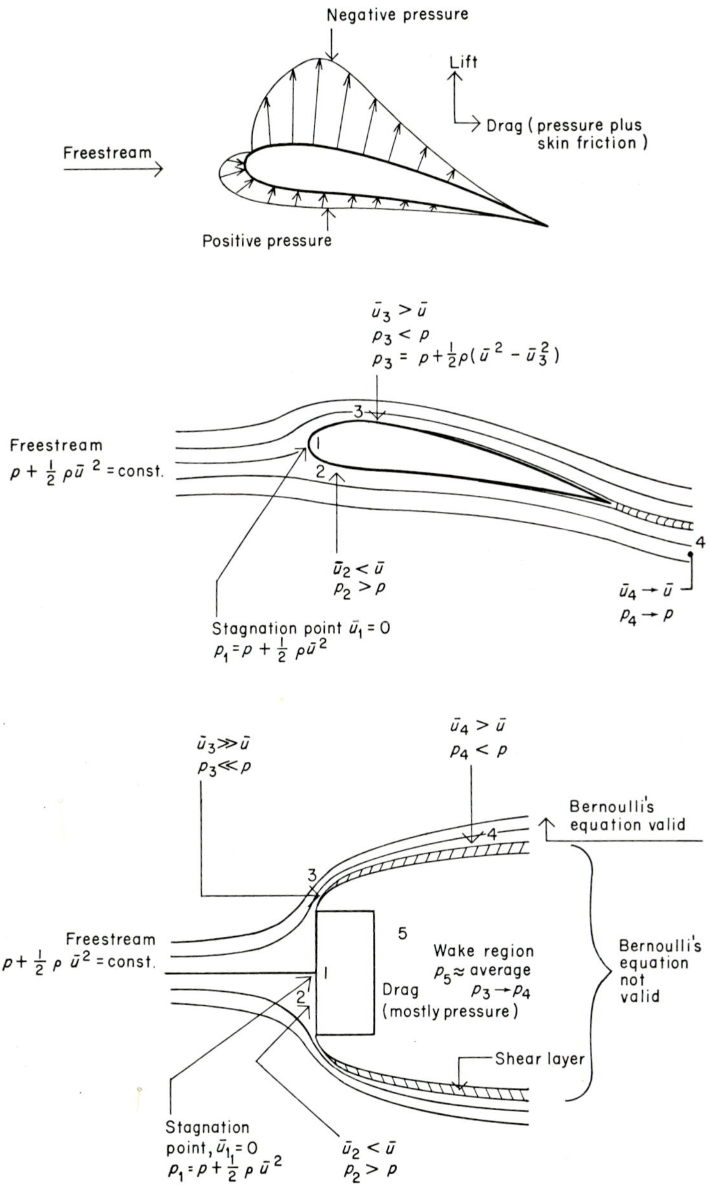

Fig. 2.5. The development of fluid pressure fields on a streamlined body and a bluff body.

The assumptions, on which the validity of Bernoulli's equation depends, effectively restrict the application to steady freestream flows and specifically exclude shear and boundary layer flows where viscous effects are significant and elemental rotation occurs and wake flows where rotation also occurs over a range of scales. However, the curvature of a surface or a shear layer may distort the freestream flow causing local changes in velocity which in turn cause changes in pressure. These pressure fields when transmitted through a shear layer and/or wake onto the surface of a solid body contribute to the normal pressure forces such as lift and drag, on that body. Examples of these pressure fields and how they are generated through the mechanism of Bernoulli's equation are given in Fig. 2.5.

2.4. The Venturi Meter and Pitot-static Tube

Measurement of flow velocities are fundamental to any study in aerodynamics. The venturi meter and pitot tube are flow measuring devices which depend on Bernoulli's equation and so it is appropriate to introduce them at this stage.

The venturi meter is used for measuring flow in a pipe and it does so by reducing the cross-section area of flow and comparing the static pressures measured before and at the reduction as shown diagrammatically in Fig. 2.6. The static pressure differences can be measured directly by a fluid manometer.

Applying Bernoulli's equation between 1 and 2 gives

$$p_1 + \tfrac{1}{2}\rho \bar{u}_1{}^2 = p_2 + \tfrac{1}{2}\rho \bar{u}_2{}^2 \tag{2.7}$$

and from continuity,

$$\bar{u}_1 A_1 = \bar{u}_2 A_2 \tag{2.8}$$

For the manometer, fluid static relationships give,

$$p_1 + \rho g z_1 = p_2 + \rho g (z_1 - h) + \rho_M g h \tag{2.9}$$

Combining eqns. (2.7), (2.8) and (2.9) gives an expression for the volume flow rate

$$Q = u_2 A_2 = A_2 \left(\frac{2gh(\rho_M - \rho)}{\rho\left(1 - \left[\frac{A_2}{A_1}\right]^2\right)} \right)^{\frac{1}{2}} \tag{2.10}$$

Flow in a tube does not behave completely in an inviscid manner, and in particular there are boundary layers, the properties of which will be shown later to depend on a parameter called the Reynolds

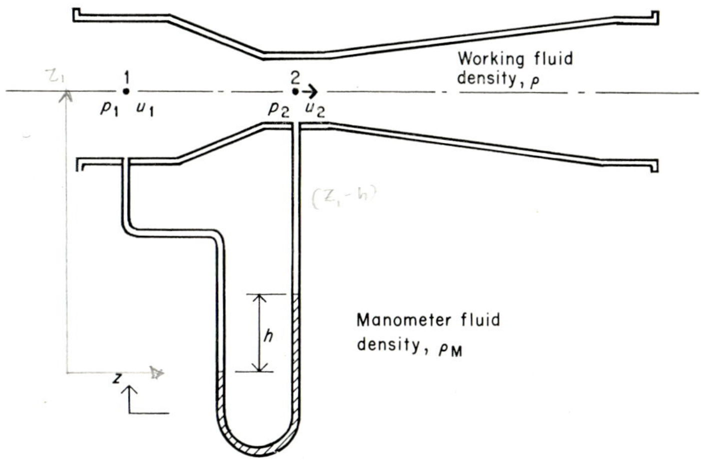

Fig. 2.6. Diagrammatic of venturi meter and U-tube manometer.

number. To account for the inviscid behaviour a correction coefficient, C, in introduced such that

$$Q = CA_1\left(\frac{2gh(\rho_M-\rho)}{\rho\left(1-\left[\frac{A_2}{A_1}\right]^2\right)}\right)^{\frac{1}{2}} \tag{2.11}$$

The value of C is between 0·95 and 0·99 for Reynolds numbers ($R_e = \rho\bar{u}D/\mu$, based on inlet conditions at 1) greater than 10^4; reference should be made to detailed texts or tables on flow measurement [2.1] if accurate values of the coefficient, C, are required.

The pitot-static tube is a device which can be inserted into a flow to measure the static and total pressure as shown set up in Fig. 2.7. Consider the streamline from 1 in the freestream to 2 just in the entrance of the total pressure measuring tube. This is called the total or stagnation pressure because ideally it is the pressure which results when the velocity is brought to zero and all the kinetic energy has been converted into pressure energy. From Bernoulli's equation

$$p_1 \pm \tfrac{1}{2}\rho\bar{u}_1{}^2 = p_2 + 0 \qquad (2.12)$$

giving

$$\bar{u}_1 = \left(\frac{2(p_2 - p_1)}{\rho}\right)^{\frac{1}{2}} \qquad (2.13)$$

the difference between the static pressure p_1 and total pressure p_2 is called the dynamic pressure, q, i.e. total pressure − static pressure = dynamic pressure.

$$p_2 - p_1 = \tfrac{1}{2}\rho\bar{u}^2 = q \qquad (2.14)$$

The dynamic pressure can be measured directly using a manometer as shown in Fig. 2.7, and inserting the fluid static relations in eqn. (2.13) gives,

$$\bar{u}_1 = \left(\frac{2gh(\rho_M - \rho)}{\rho}\right)^{\frac{1}{2}} \qquad (2.15)$$

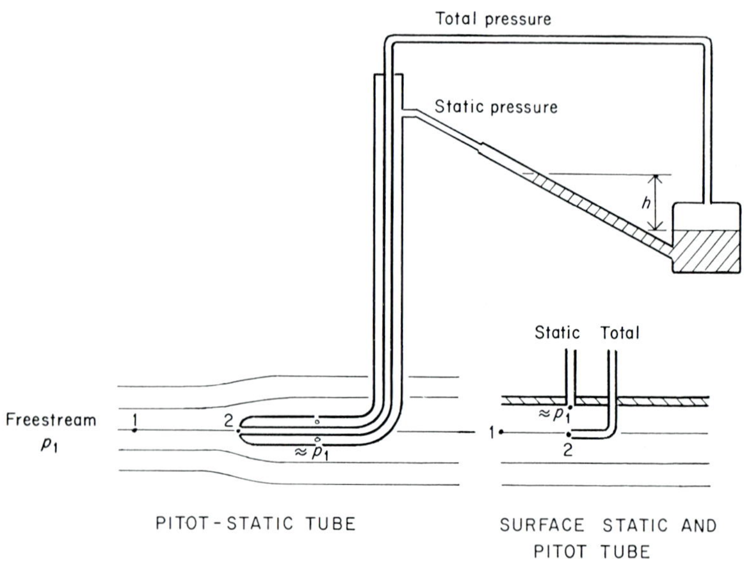

Fig. 2.7. A pitot-static tube connected to a sloping tube manometer.

The total and static pressures near a solid boundary can also be simply measured using a surface static vent and a pitot tube bent to face the stream which measures total pressure as shown in Fig. 2.7.

The earlier types of pitot-static tubes also required a significant correction coefficient to allow for viscous effects and errors incurred by the position of the static pressure measuring vent. However, since the early 1960s, pitot-static tubes with near ellipsoidal heads and carefully positioned static vents have been developed, principally at the National Physical Laboratory, U.K. [2.2] which have reduced the correction coefficient to between 0·99 and 1·01 for a large range of Reynolds numbers. These types of pitot-static tube are readily available on the commercial market and with these the use of a correction coefficient has become virtually redundant.

2.5. Energy Considerations in Flow Systems

The terms in Bernoulli's equation reflect a statement of the First Law of Thermodynamics, that energy is conserved. However, when discussing losses in duct and ventilation systems, it is essential that a control volume be defined and the energy equation be used in its correct sense.

2.5.1. The Energy Equation

Any fluid system can be surrounded by a control volume across the boundaries of which an energy conservation statement in terms of gains and losses can be written. For the design of ducts or air transfer systems for ventilation and air conditioning applications it is usually satisfactory to assume that the air flow is steady and isothermal, although practice may be far from this on occasions. This effectively separates the energy considerations associated with the provision of heating and cooling from the energy considerations necessary for the design of the air transfer system. The calculation of heat load on a building for the design of air conditioning systems is outside the scope of this text and has been covered earlier in this Architectural Science Series by Van Straaten [2.3].

There are a number of ways of writing the energy equation, but for duct systems it is conventionally written in terms of energy per unit volume. This has the dimensions of pressure, i.e. dimensionally,

$$\frac{\text{energy}}{\text{volume}} = \frac{ML^2t^{-2}}{L^3}$$

$$= \frac{MLt^{-2}}{L^2}$$

$$= \frac{\text{force}}{\text{area}} = \text{pressure}$$

The components of the energy equation in terms of energy per unit volume are derived from a number of sources as follows:

(a) stored energy, kinetic energy, $\frac{1}{2}\rho v^2$, potential energy, ρgz, internal energy, which in the absence of chemical changes, remains unchanged in an isothermal system;
(b) flow energy, pressure energy, p;
(c) mechanical and viscous energy gain and losses, energy gain, p_g, energy loss, p_l.

With reference to the control volume and variables defined in Fig. 2.8 the energy in can be equated to the energy out as follows:

$$p_1+\tfrac{1}{2}\rho_1\bar{u}_1{}^2+\rho_1 gz_1+p_g = p_2+\tfrac{1}{2}\rho_2\bar{u}_2{}^2+\rho_2 gz_2+p_l \qquad (2.16)$$

For the design of duct and ventilating systems, where the working fluid is air, it is usually satisfactory to regard the effect of the potential energy terms and the effect of air density charges as being negligibly small.

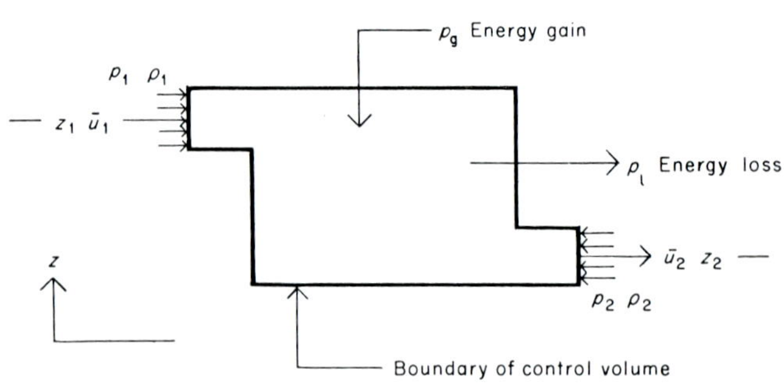

Fig. 2.8. Control volume for energy considerations.

The pressure gain in the system is usually provided by a fan and can be given by

$$p_g = \frac{\eta P}{Q} \tag{2.17}$$

where P = fan shaft power, η = fan efficiency, Q = volume flow rate.

The pressure losses in a system can be due to duct friction, flow losses across heating or cooling coils, bends, control valves and registers, etc. The total loss can be obtained by summing all the component losses which can be determined for each in the form

$$p_l = C\tfrac{1}{2}\rho\bar{u}^2 \tag{2.18}$$

where $\bar{u}$ is the local mean velocity, and C is the pressure drop coefficient for the component which can be obtained from manufacturers' tables, handbooks, data sheets and texts [2.4].

In some cases the energy equation, including gain and loss terms, is quoted in terms of pressure head. This means that pressure is replaced by the head of water which would produce that pressure and is simply achieved by dividing all terms in eqns. (2.16), (2.17) and (2.18) by $\rho_{water}g$. The dimension of these equations is then length (i.e. head of water) and in fact represents energy per unit weight.

2.5.2. *A Duct System*

A typical duct system is shown in Fig. 2.9, in which the heating

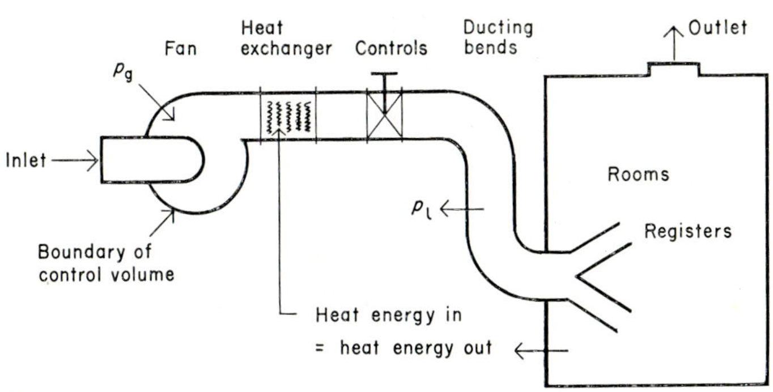

Fig. 2.9. Control volume for an air conditioning duct system.

losses and gains are shown to be equal and hence do not contribute directly to the duct system design. The control volume in this case, as shown, includes all the duct system and the rooms.

In particular, the inlet and outlet conditions may be such that pressure differences are negligibly small (although if they were on opposite sides of a building, in strong wind conditions, this would not be true) and the inlet and outlet velocities may be similar or negligibly small in which case the energy eqn. (2.16) reduces to

$$p_g = p_l \tag{2.19}$$

2.5.3. *Free Ventilation*

The energy equation is also used when calculating natural or free ventilation through a house where the pressure drop due to the wind provides the energy input and the resistance in the control volume is provided by the flow through windows, louvres, screens and internal openings as shown in Fig. 2.10.

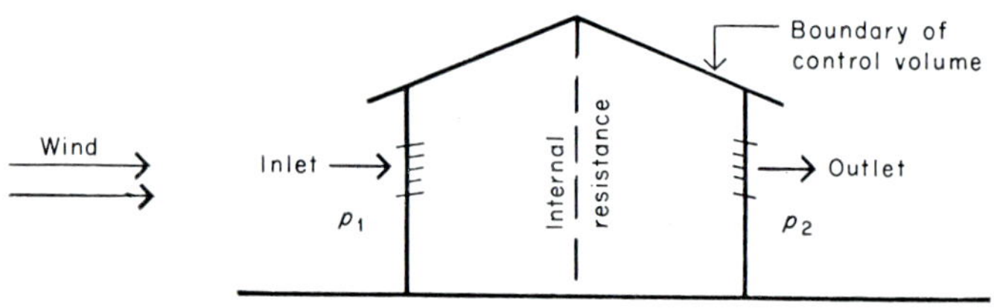

Fig. 2.10. Natural ventilation of a building.

In this case, as inlet and outlet velocities are likely to be similar, the energy eqn. (2.16) reduces to

$$p_1 - p_2 = p_l \tag{2.20}$$

2.5.4. *Stack Effect*

The 'stack effect' or principle on which a chimney stack draws air is a case where density differences are not only significant but are essential for the process to operate. Basically, the air in a stack is heated and as a consequence of having a lower density than the outside air, it rises and so causes a draught up the chimney. The same thermal or convection process is utilised by ridge and high ventilators to promote air circulation through buildings on calm, hot days. The fluid mechanics of the stack effect can be explained with reference to Fig. 2.11.

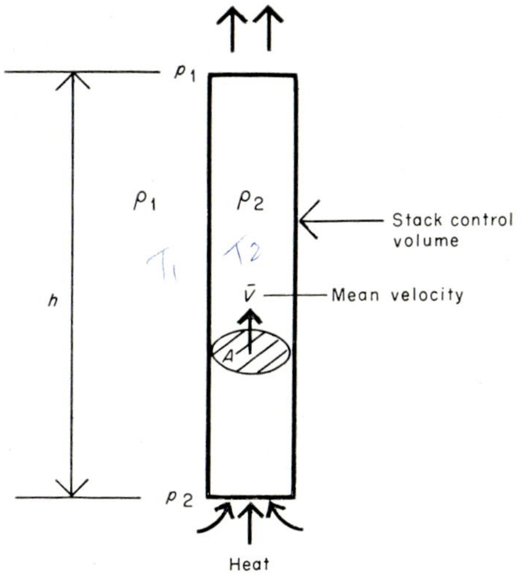

Fig. 2.11. Simplified representation of the stack effect.

The analysis of the stack effect does not directly involve the energy equation and in fact must be set up by using momentum considerations, that is, the vector sum of the forces on the fluid in a control volume equals the rate change of momentum. In the simplified case given in Fig. 2.11, a control volume has been set up inside a parallel sided stack of height h, and cross-sectional area A, in which the air is being heated such that it has an average density ρ_2, whilst the outside air has a density ρ_1. The mean velocity in and out of the stack, $\overline{V}$, in Fig. 2.11 is the same, hence the rate change of momentum is zero, and the sum of the forces on the fluid in the control volume resolved vertically is zero, which gives

$$p_2A - p_1A - F - \rho_2 ghA = 0 \tag{2.21}$$

where F is defined as the sum of all the internal friction and resistance forces caused by the stack on the air flowing up the control volume.

Hydrostatic relationships give,

$$p_2 = p_1 + \rho_1 gh \tag{2.22}$$

which on substitution into eqn. (2.21) gives,

$$F = ghA(\rho_1 - \rho_2) \tag{2.23}$$

Now the force F can be expressed in terms of a pressure loss coefficient C, based on the area A, and velocity $\overline{V}$, as defined in eqn. (2.18),

$$F = C\tfrac{1}{2}\rho_2 \overline{V}^2 A \tag{2.24}$$

which on substitution into eqn. (2.23) gives,

$$\overline{V} = \left(\frac{2gh}{C}\frac{(\rho_1 - \rho_2)}{\rho_2}\right)^{\frac{1}{2}} \tag{2.25}$$

From the perfect gas equation,

$$p/\rho = RT \tag{2.26}$$

where R is the gas constant and T is the absolute temperature, eqn. (2.25) can be written in terms of temperature rather than density difference as

$$\overline{V} = \left(\frac{2gh}{C}\frac{(T_2 - T_1)}{T_1}\right)^{\frac{1}{2}} \tag{2.27}$$

The quantity of air flow due to the stack effect is sometimes quoted with a single constant as,

$$Q = CA[H(T_2 - T_1)]^{\frac{1}{2}} \tag{2.28}$$

where the constant C embodies the coefficients of resistance for the system, the ambient temperature and gravitational acceleration, and H is the average height between inlet and outlet.

2.6. Resistance to Fluid Motion

2.6.1. *Viscosity*

A fluid can be defined as a substance that deforms continuously when subjected to a shear stress. Viscosity is the basic fluid property which provides resistance to rate of shear. This concept may be simply explained with reference to an element a, b, c, d in a fluid field between two parallel plates, a distance, s, apart which are moving with a constant velocity, u, relative to each other as shown in Fig. 2.12.

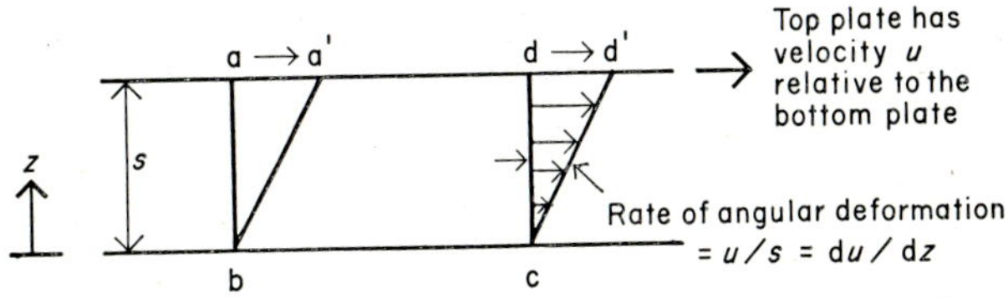

Fig. 2.12. Deformation of an element of fluid due to a constant shear force.

The rate of deformation, u/s, can be expressed as a velocity gradient in differential form as $\mathrm{d}u/\mathrm{d}z$ and if the intensity of shear, that is force resisting the motion per unit area, is called the shear stress, τ, these two can be related by a constant of proportionality

$$\mu = \frac{\tau}{\mathrm{d}u/\mathrm{d}z} \qquad (2.29)$$

where μ is called the dynamic viscosity. If other variables such as temperature and pressure are constant, then it has been found that for fluids such as air and water μ is a true constant and these fluids are known as Newtonian fluids which follow 'Newton's Law of Viscosity' normally expressed as

$$\tau = \mu \frac{\mathrm{d}u}{\mathrm{d}z} \qquad (2.30)$$

There are substances which, whilst behaving as fluids, are non-Newtonian, because they exhibit either temporary elastic properties, such as bitumen, or resistance to shear, which does not vary linearly with rate of shear, such as some of the long-chained hydrocarbons used in paint.

The mechanism of viscosity is different for liquids and gases. For liquids viscosity has its origin in the cohesion between molecules which decreases with increasing temperature. For gases, where the molecules are much further apart, the viscous action is not only relatively less, but is attributed to a momentum transfer process as molecules cross from one moving layer to the next, and as molecular activity increases with temperature, so does the viscosity of a gas increase with temperature. Values of dynamic viscosity (and density) for air and water are given in Appendix I.

2.6.2. Drag

The total force, in the streamwise direction generated by the action of a fluid stream on an immersed body is called the 'drag' force.

The drag force in subsonic aerodynamics is generally recognised to be made up of contributions from the action of viscosity at the surface, termed 'friction drag' and the streamwise component of the pressure distribution on the body termed the 'pressure drag'. The drag of slender bodies in the streamwise direction, such as aeroplane wing or fuselage, is mostly made up of friction drag. As bodies become thicker in the direction normal to the fluid stream, the contribution from pressure drag increases. At a certain thickness (relative to depth) the flow separates and such a body would generally be called a bluff body, the drag of which would be almost entirely due to pressure drag. Pressure drag is relatively very much greater than friction drag, hence the need to streamline racing cars and aeroplanes to reduce pressure drag. This can be simply illustrated by noting that at similar, high wind speeds the drag of an aerofoil with a thickness to chord ratio of 12% has approximately 1/50 of the drag of a circular cylinder of the same thickness.

In architectural aerodynamics we are mostly dealing with bluff bodies with high pressure drag components as illustrated in Fig. 2.5. However, it will be shown in later sections that the effects of viscosity are very significant in determining the flow fields and hence the pressure fields from which the pressure drag is derived.

2.7. Dimensional Analysis, Scaling and some Important Coefficients

When a phenomenon, or process, involves a number of variables, such as length, velocity, density, viscosity, pressure and so on, it becomes very clumsy to discuss in terms of the effect of each variable. Fortunately, it is not necessary to do this because by a procedure of dimensional analysis it is possible to formulate a relation between sets of dimensionless groups of variables which number less than the dependent variables. In fact, a theorem developed by Buckingham [2.5] defines the degree of simplification possible as follows:

> The smallest number of independent dimensionless groups that may be employed to describe a phenomenon known to involve n variables is equal to the number $n-r$, where r is the number of basic dimensions needed to express the variables dimensionally.

Most of the important dimensionless groups used to simplify the presentation of data in the aerodynamics field are in fact derivatives of force ratios which are used to define model scaling requirements. Hence it is worth briefly developing these scaling requirements. Similarity in scaling an event or process from one scale to another can be achieved in varying degrees from

> geometric similarity, where ratios of linear dimensions are equal; to, kinematic similarity, where particle paths are geometrically similar; to, dynamic similarity, where the ratios of forces are equal.

In fluid mechanics it is virtually essential to achieve dynamic similarity for the significant forces in any scaling exercise, or to admit corrections for any inability to achieve dynamic similarity. Dynamic similarity exists between geometrically and kinematically similar systems 1 and 2 if the ratio of all the forces are equal,

$$\text{e.g.}\quad \frac{(\text{pressure force})_1}{(\text{pressure force})_2} = \frac{(\text{viscous force})_1}{(\text{viscous force})_2} = \frac{(\text{inertia force})_1}{(\text{inertia force})_2} \tag{2.31}$$

and so on as illustrated in Fig. 2.13.

Equation (2.31) may be written in the form

$$\frac{(\text{pressure force})_1}{(\text{inertia force})_1} = \frac{(\text{pressure force})_2}{(\text{inertia force})_2} = \text{Constant}$$

and

$$\frac{(\text{viscous force})_1}{(\text{inertia force})_1} = \frac{(\text{viscous force})_2}{(\text{inertia force})_2} = \text{Constant} \tag{2.32}$$

Table 2.1
Some dimensionless coefficients used in aerodynamics to simplify data presentation and as scaling criteria

1. $$\frac{\text{pressure force}}{\text{inertia force}} \quad \frac{pA}{Ma} = \frac{pL^2}{\rho L^2 V^2}$$

Pressure coefficient, $C_p = \dfrac{p}{\frac{1}{2}\rho V^2}$

Lift coefficient, $C_L = \dfrac{\text{Lift}}{\frac{1}{2}\rho V^2 A}$

Drag coefficient, $C_D = \dfrac{\text{Drag}}{\frac{1}{2}\rho V^2 A}$

Moment coefficient, $C_M = \dfrac{\text{Moment}}{\frac{1}{2}\rho V^2 LA}$

2. $$\frac{\text{inertia force}}{\text{viscous force}} \quad \frac{Ma}{\tau A} = \frac{Ma}{\mu\left(\frac{dV}{dy}\right)A} = \frac{\rho L^2 V^2}{\mu\left(\frac{V}{L}\right)L^2}$$

Reynolds number $Re = \dfrac{\rho VL}{\mu}$

3. $$\frac{\text{inertia force}}{\text{gravity force}} \quad \frac{Ma}{Mg} = \frac{\rho L^2 V^2}{\rho L^3 g}$$

Froude number, $F = \dfrac{V}{(Lg)^{\frac{1}{2}}}$

4. $$\frac{\text{inertia force}}{\text{elastic force}} \quad \frac{Ma}{KA} = \frac{\rho L^2 V^2}{KL^2}$$

Stiffness number (Cauchy number) $= \dfrac{\rho V^2}{K}$ or $\dfrac{\rho V^2}{E}$

Mach number $M = \dfrac{V}{(K/\rho)^{\frac{1}{2}}} = \dfrac{V}{C}$

Notes
1. Pressure p must be referred to some reference, usually freestream static pressure.
2. $\frac{1}{2}\rho V^2$ is conventionally used in these dimensionless numbers because it is conveniently obtained as dynamic pressure.
3. Reference length L, and area A must be defined for the specific application. For the Reynolds number and drag of bluff bodies L is the width normal to the stream and A is the projected area normal to the stream direction. Similarly, for streamlined bodies L is the streamwise chord length and A is the planform area. For duct flows the Reynolds number is based on diameter or equivalent hydraulic diameter if the duct is not circular.

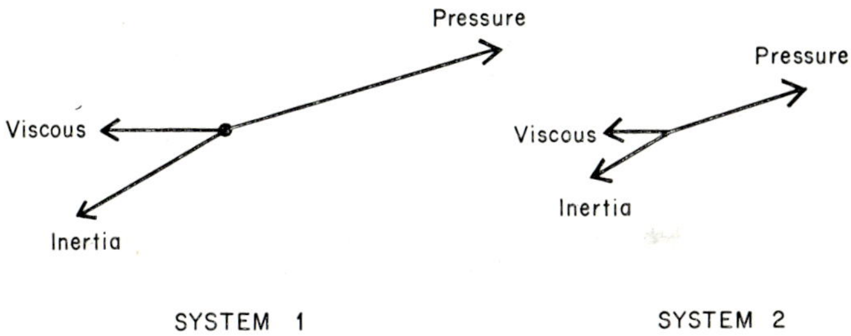

Fig. 2.13. Dynamic similarity between systems requires that the ratio of forces be equal.

Table 2.2

Some homogeneous units used in fluid mechanics

System	Basic units			
	Length	Mass	Time	Thermodynamic temperature
Système International (Absolute MKS)	metre m	kilogram kg	second s	kelvin degree K
English—Type 1 (Engineering or technical)	foot ft	slug slug	second s	rankine degree R

System	Derived units						
	Force	Work, energy, quantity of heat	Power	Density	Velocity	Pressure	Viscosity (dynamic absolute)
Système International (Absolute MKS)	newton N = kg m/s²	joule J = N m	watt W = J/s	kg/m³	m/s	pascal N/m²	Ns/m² or kg/m sec
English—Type 1 (Engineering or technical)	pound force lbf = slug ft/sec²	ft lbf	ft lbf/sec	slug/ft³	ft/sec	lbf/ft²	lbf sec/ft or slug/ft sec

It has been found convenient to express the scaling force ratios with reference to the inertia force and these ratios form the basis of the commonly used dimensionless coefficients which are listed in Table 2.1.

It should be noted that the inertia force ratio between two systems can be expressed as a ratio of the variables as follows:

$$F_r = \frac{(\text{inertia force})_1}{(\text{inertia force})_2} = \frac{(Ma)_1}{(Ma)_2}$$

$$= \frac{(\rho L^3 V^2 L^{-1})_1}{(\rho L^3 V^2 L^{-1})_2}$$

$$= \frac{(\rho L^2 V^2)_1}{(\rho L^2 V^2)_2} = \rho_r L_r^{\,2} V_r^{\,2} \quad (2.33)$$

where M = mass, a = acceleration, L = length, V = velocity, ρ = density, and subscript r is used to describe the ratio between similar variables in the systems 1 and 2.

In any presentation or reduction of data, it is essential that a homogeneous (consistent) system of units be used. In general, the Système International (SI) is being used or phased in throughout the world, but there is still some reliance on the English Engineering system of units. As a reference Table 2.2 has been prepared to give the basic and some commonly used derived units in both systems.

2.8. The Boundary Layer and Separation

The boundary layer is a shear layer formed by the action of shear stresses at a solid boundary. It is a relatively thin layer with a mean velocity profile starting at zero at the solid boundary and increasing to the freestream velocity at the outer edge. The boundary layer is dominated by the effects of viscosity and there are two very distinct types of boundary layer the presence of which can be predicted with relation to the Reynolds number introduced in Section 2.7 as the ratio of inertia to viscous forces. At low values of Reynolds numbers ($Rex = \rho Vx/\mu$, where x is the distance from the start), typically less than 10^5, such as a knife being drawn through honey, the boundary layer is laminar, that is it is characterised by stratified layers of fluid sliding over each other with very little mixing across the layer. At higher Reynolds numbers the boundary layer first becomes unstable with bursts of turbulence and a transition occurs followed by what is called a turbulent boundary layer as illustrated in Fig. 2.14.

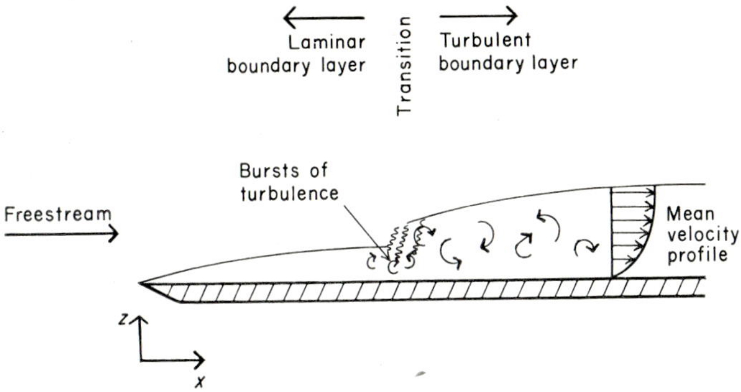

Fig. 2.14. Transition from a laminar to a turbulent boundary layer.

The turbulent boundary layer is characterised by turbulent eddies causing considerable mixing through the layer such that the losses are mainly due to viscosity associated with the dissipation of the eddies rather than the direct shearing action as in the laminar boundary layer. A comparison of the characteristics of the two boundary layers is given in Table 2.3. It should be noted that most architectural aerodynamic flows occur at relatively high Reynolds numbers and as such the boundary layers are turbulent although the effects of transition on separation from curved surfaces will be seen to be important. It is also important to realise that the wind near the earth's surface is in fact just a rough wall, turbulent boundary layer, several hundred metres thick.

The formulation of the boundary layer theory by Prandtl (1904), which laid the basis for modern fluid mechanics, was to some extent brought on by his studies of flow separation in diffusers. He stated that the boundary layer separation depended on the presence of an increasing (adverse) pressure gradient, in a streamwise direction, along the boundary. Much subsequent research has supported this

Table 2.3

Comparative properties of boundary layers

Property	Laminar boundary layer	Turbulent boundary layer
Thickness	Thin, grows slowly	Thick
Composition	Stratified flow	Turbulent eddies with a range of sizes
Velocity distribution	$\frac{u}{z} = \frac{4{\cdot}9}{(Rex)^{\frac{1}{2}}}$	Smooth wall $\frac{u}{z} = \left(\frac{z}{8}\right)^{1/7}$ Rough wall $\frac{u}{z} = \left(\frac{z}{8}\right)^{1/7 \rightarrow 1/2}$ Turbulence greater over the lower part
Reynolds number	Exists only at low values ($Rex < 10^5$)	Exists at any but very low values
Surface friction	Low	High
Stability	Low, minor instabilities cause transition and separation	High, due to momentum transfer capability of turbulent motion

Factors affecting transition

Reynolds number, stream turbulence, heat transfer, pressure gradient, surface roughness, noise.

finding and the mechanism can be explained with reference to Fig. 2.15.

In a decelerating flow the retarded flow in the boundary layer, as it stretches in thickness, suffers a relatively greater deceleration than the freestream flow. The momentum of the flow near the solid boundary is small and eventually reaches the state where it can no longer oppose the adverse pressure gradient imposed by the free-stream, and reverse flow eventually occurs. This reverse flow at the base of the boundary layer means the boundary layer has been separated from the solid boundary and is now a free shear layer. It can now be seen why the turbulent boundary layer is relatively more stable because the turbulent eddies can transport momentum from the outer regions of the layer to help make up the momentum deficit occurring at the base. In this sense a turbulent boundary layer can resist separation in an adverse pressure gradient for some distance.

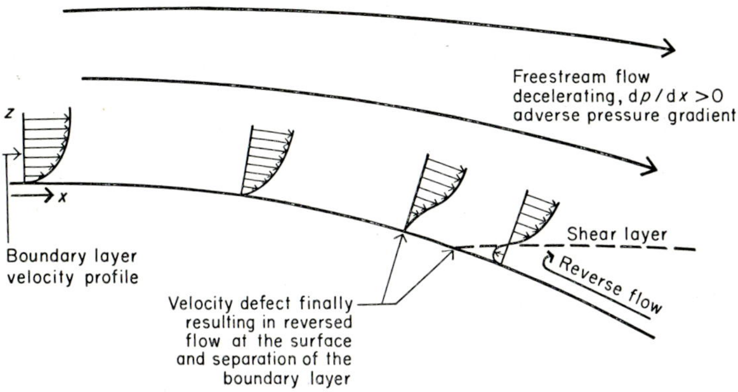

Fig. 2.15. Mean velocity profile before and through separation.

However, when a boundary layer is confronted with a sharp edge such as the corner of a rectangular section, the separation occurs immediately at the corner. Separation can be delayed or suppressed by adding momentum at the base of the boundary layer by blowing or withdrawing by sucking. Both techniques are commonly employed for boundary layer control and there are a variety of other devices such as small vortex generators which create a streamwise vortex at the edge of the boundary layer to induce freestream momentum down to the base of the boundary layer.

2.9. The Effects of Reynolds Number

The Reynolds number, coupled with the other parameters listed at the bottom of Table 2.2, is particularly important in determining the status of the boundary layer and in particular when transition and separation will occur. It is the connection with separation via the boundary layer which is most important in the development of flow fields, pressure distributions and hence aerodynamic forces on bodies. The effects of Reynolds number on the boundary layer separation points, the resultant width of the wake, base pressure and drag is probably best illustrated with the case of a circular cylinder as set out in Figs. 2.16, 2.17 and 2.18. The effect of corner radius as a circular cylinder approaches a square cylinder is also shown in Fig. 2.18. For

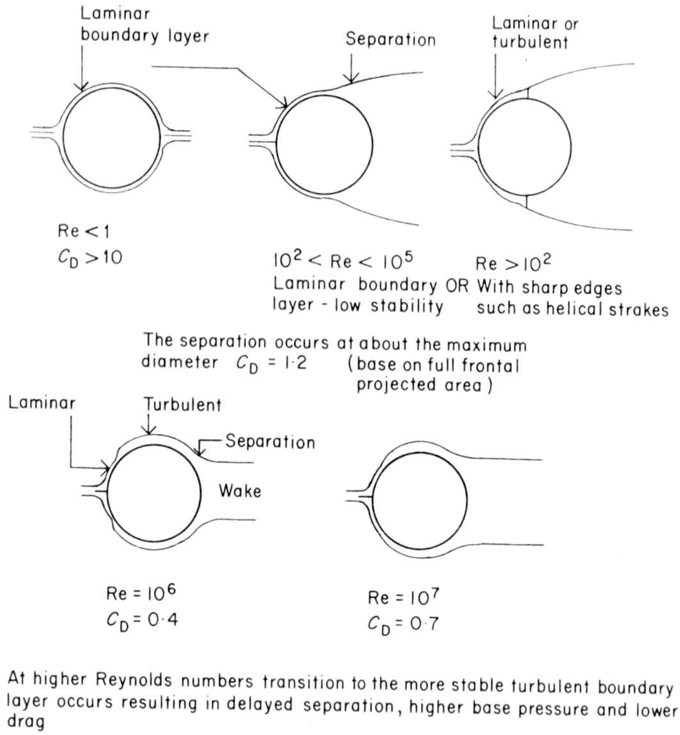

Fig. 2.16. Diagram of boundary layer separation, resultant wake widths and drag coefficients as a function of Reynolds number for a smooth two-dimensional circular cylinder in smooth flow.

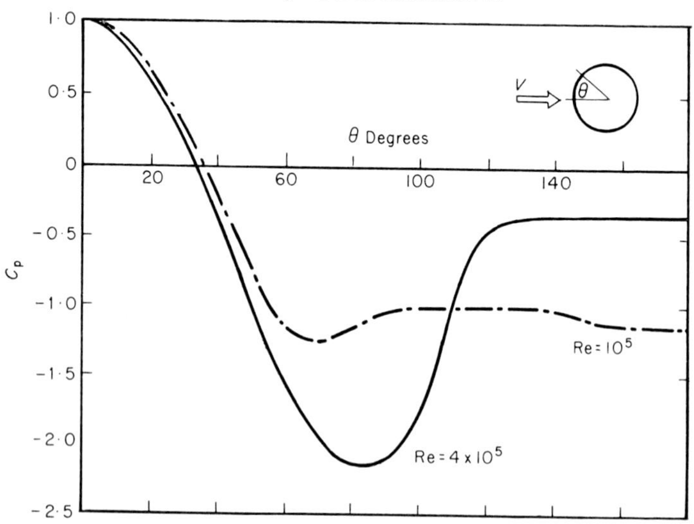

Fig. 2.17. Influence of Reynolds number on the pressure distribution around a circular cylinder (after Fage and Faulkner [2.9]).

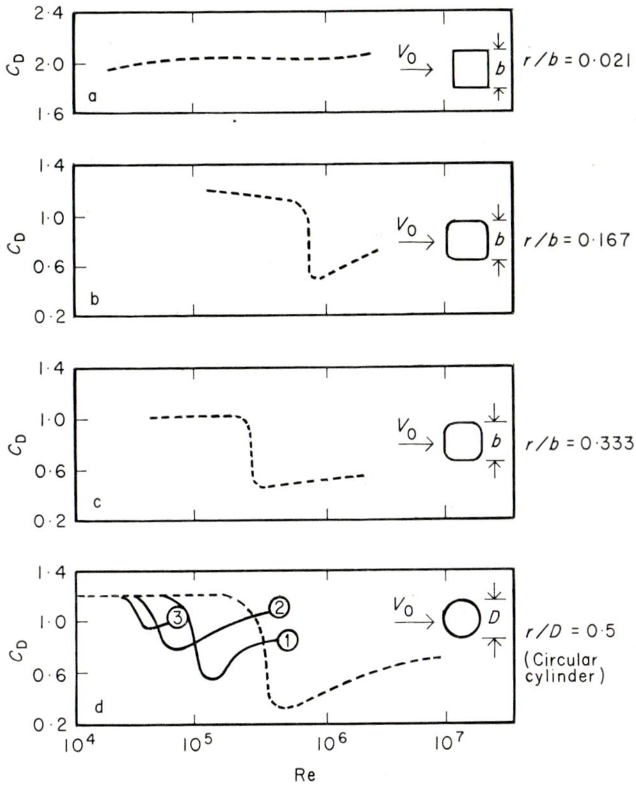

Fig. 2.18. Influence of Reynolds number, corner radius r and surface roughness on values of drag coefficient k for square and circular cylinders in smooth flow, smooth surface, sanded surface, (1) $k/D = 0{\cdot}002$, (2) $k/D = 0{\cdot}007$, (3) $k/D = 0{\cdot}020$ (after Delaney and Sorensen [2.10] and Fage and Warsap [2.11]).

the square cylinder where the separation occurs from the leading corners (with little or no fluctuating re-attachment) the effect of Reynolds number is relatively small because it is no longer the dominating factor in controlling the separation and wake width.

2.10. The Effects of Turbulence

The effects of freestream turbulence on boundary layer transition and hence separation is relatively well known. However, the effects of freestream turbulence on the shape of the shear layer shed from a bluff body (and hence drag) is only just becoming understood. It appears that this effect is very significant in the wind loading of structures, not only in terms of mean and fluctuating forces but on

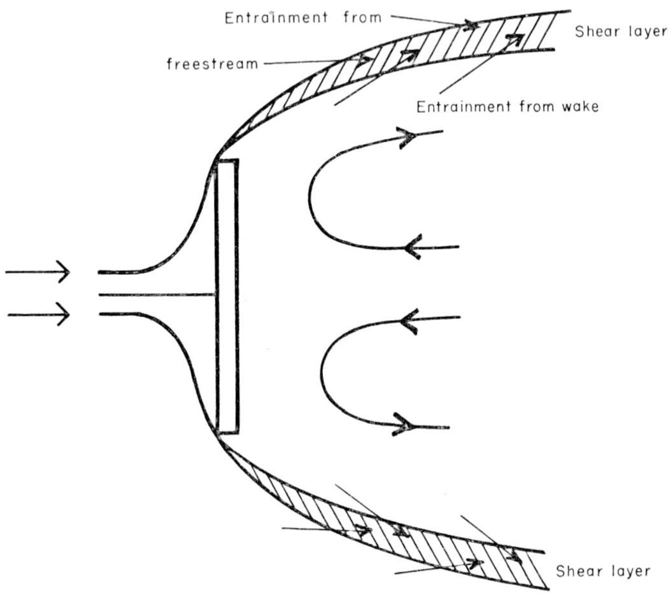

Fig. 2.19. Entrainment of freestream and wake fluid into the shear layers.

the fluctuating pressures near the leading edge which almost invariably cause the highest pressures on cladding and glazing.

The effects of freestream turbulence have been observed in several apparently paradoxical experiments. Bearman [2.6] showed how increasing freestream turbulence and turbulence length scale increased the drag of sharp edged flat plates, and Vickery [2.7] showed how increasing freestream turbulence decreased the drag of a square two-dimensional prism. The explanation was shown to lie in the behaviour of the shear layers. As noted in Section 2.3, the base pressure (i.e. wake pressure) is a result of the pressure generated by the freestream flow just outside of the shear layer. If the radius of curvature of the shear layer were to decrease, the velocity just outside would increase and via Bernoulli's equation would give a lower pressure which in turn would result in a lower base pressure and increased drag. The radius of curvature of shear layers are dependent on the rate at which freestream and wake fluid can be entrained as shown in Fig. 2.19.

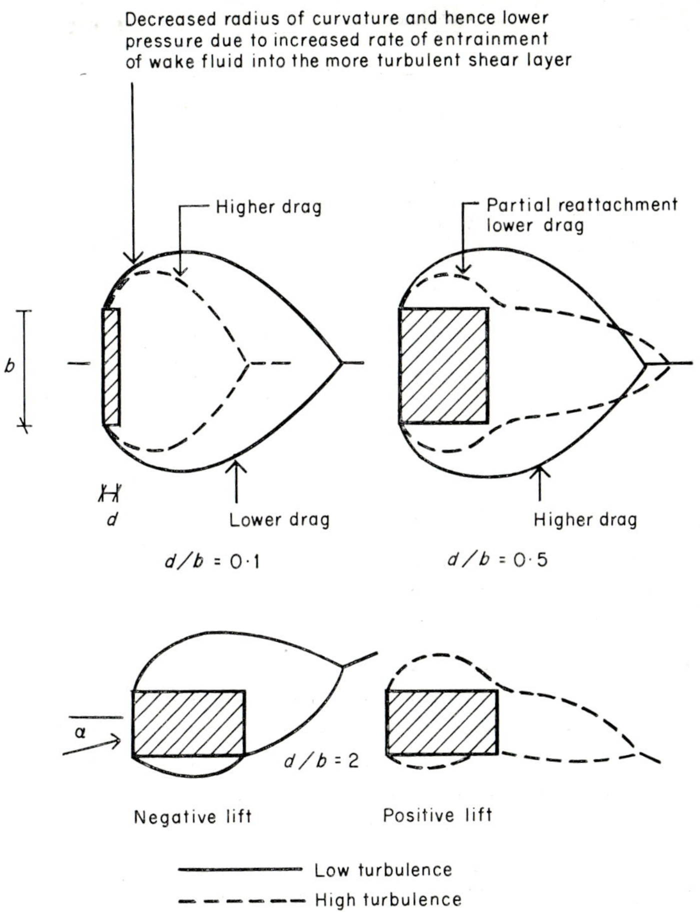

Fig. 2.20. Sketches of expected effect of turbulence on dividing streamlines as it affects drag and lift of rectangular sections (after Gartshore [2.8]).

Notwithstanding the fact that the shear layers roll up into vortices and along with the wake pressures are continually fluctuating, a description of the effect of freestream turbulence will be given as though the shear layers are steady. The effect of increasing turbulence is to increase the rate of entrainment of fluid into the shear layer,

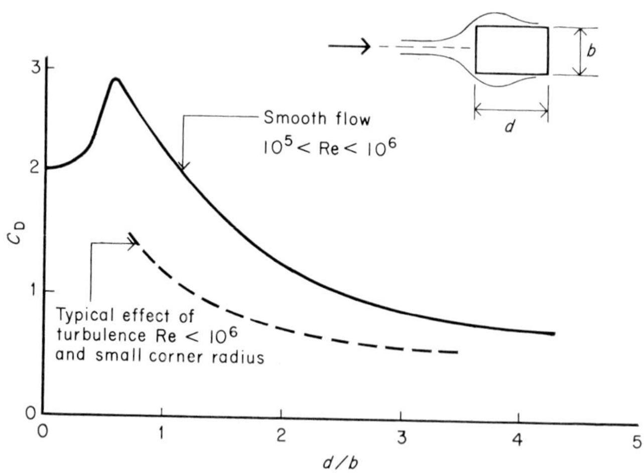

Fig. 2.21. Effect of turbulence and after-body length on the drag of two-dimensional cylinders (after Bearman [2.6], Delaney and Sorenson [2.10] and Vickery [2.7]).

which in turn reduces the radius of curvature. In the case of a flat plate the reduced radius of curvature of the shear layer leads to a reduced base pressure and increased drag, but for a square section, for example, it leads to re-attachment of the shear layer with reduced pressure on the streamwise faces but increased base pressure and hence lower drag. These effects are described diagrammatically in Fig. 2.20. Gartshore [2.8] took this explanation a step further when he showed that it was only the fine scale turbulence that was responsible for controlling the rate of entrainment into the shear layer and he reproduced the previously described effects by using the wake of a thin wire upstream of the cylinder in smooth flow.

The effect of turbulence and after-body length on the drag of a rectangular cylinder is illustrated in Fig. 2.21.

Another most important aspect of freestream turbulence is the effect it has on periodicity in the wake. As previously noted, the shear layers actually roll up into vortices which are unsteady and continually fluctuating. In the case of smooth flow past a bluff cylinder, the vortices shed alternately and periodically from either side of the cylinder, forming what is called the Von Karman Vortex Street. As the freestream flow becomes more turbulent, the very orderly periodic shedding of vortices starts to break down until eventually the wake contains a broad range of shedding frequencies. The cross-wind oscillations of tall and slender structures tend to be dominated by the wake excitation mechanism. This is particularly evident with tall chimney stacks which, if they do not have sufficient inherent damping, will oscillate to large amplitudes at wind speeds where the dominant vortex shedding frequency coincides with the natural frequency of the stack.

2.11. Wind Tunnel Testing

In later chapters references will be made to wind tunnel tests in almost every area. Hence an introduction to wind tunnels and testing techniques will be set out here.

For many years people studying architectural aerodynamics used aeronautical type wind tunnels with smooth flow characteristics (low turbulence). Similarly, some flow visualisation exercises were carried out in smoke tunnels also using smooth flow. However, the natural wind is turbulent and since about 1960 evidence has mounted steadily to show that it is essential to model the turbulence of the natural wind if meaningful measurements and observations are to be made in wind tunnel studies.

Techniques for the modelling of wind effects on buildings and structures have improved considerably in recent years with the advent of several large wind tunnels designed to produce turbulent boundary layer models of the natural wind. The modelling requirements to measure wind effects are generally agreed amongst workers in the field. However, for a number of reasons the techniques used and assumptions made to achieve these requirements vary considerably. Probably the main reason for this is not so much lack of agreement but variations in the capabilities of the various wind tunnel facilities and associated equipment.

2.11.1. Types of Wind Tunnel Testing

Wind tunnel tests currently being conducted on buildings and structures can be divided into two types. The first is concerned with the determination of wind loading effects to enable the design of a wind resistant structure. The second is concerned with the flow fields induced around the structure, for example as it affects pedestrian comfort and safety at ground level, or the rise of a plume from a chimney stack, and concentration levels of exhaust pollutants.

The type of tests used to determine wind loading effects can be divided further into those which are quasi-steady and those which model the dynamics of both the wind and the structure. The quasi-steady type of tests usually involve the measurement of mean (sometimes called static) pressure distributions or force on a structure. The pressure or force coefficients so determined are then used to calculate full scale loadings via a maximum gust wind speed usually called the design wind speed. This approach is relatively simple and quite satisfactory for many structures and is the approach used in most of the wind loading codes. The limitations of the quasi-steady approach occur when the wind loading on the structure is significantly increased or decreased by the response of the structure to the energy available at resonant frequencies for both along-wind and cross-wind excitations. When this occurs, a full dynamic test is usually required to determine the wind loading. The quasi-steady approach is also very limited when determining peak local pressures, which more often than not are intermittent and the result of fluid instabilities. For these studies it is essential that the velocity profile and turbulence characteristics of the natural wind be modelled correctly.

Tests for modelling flow fields and diffusion rates also generally require correct modelling of the velocity profile and turbulent properties of the natural wind. The velocity gradient is most important as the upstream face pressure gradients and hence flow fields are directly related to the local incident dynamic pressure. In a few rare cases the turbulent properties are not quite so important in detail, as for example the entrainment and hence diffusion rate of a buoyant plume which in the initial phase is more dependent on internal plume turbulence which is dominated by buoyant motions.

2.11.2. *Natural Wind Models*

The modelling of wind effects on structures first requires that a satisfactory model of the natural wind be produced and then that a similarly scaled structural model be tested in this model wind environment. Criteria and techniques for achieving a model of the

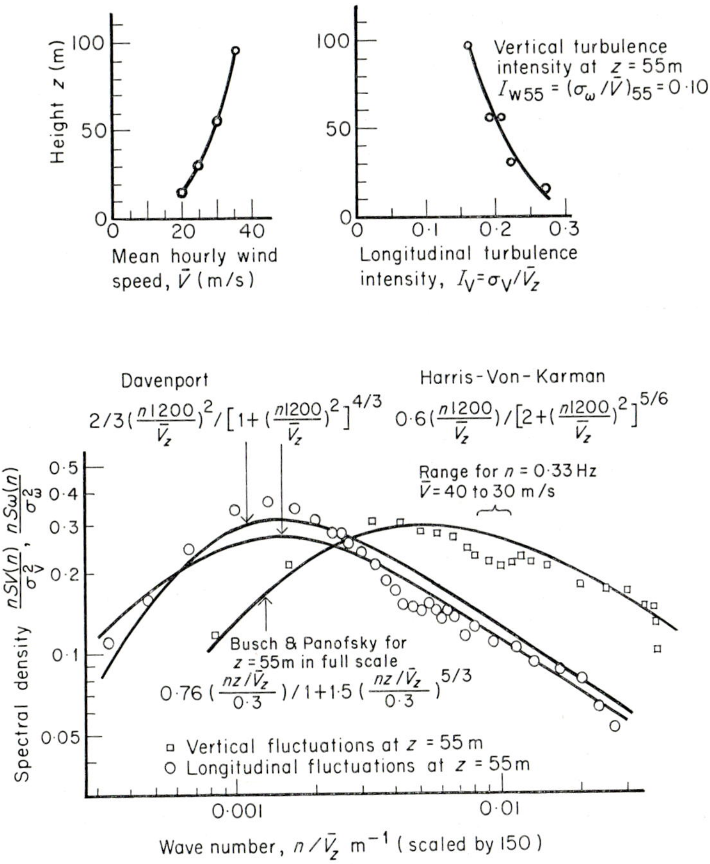

Fig. 2.22. Characteristics of 1/150 scale model of wind flow over suburban-factory fetch in the Monash University 4×3 m wind tunnel.

natural wind have been discussed by many authors; reference can be made to the proceedings of the four International Conferences on Wind Effects on Buildings and Structures referred to in Chapter 1.

If the natural wind is to be modelled at all accurately, the minimum requirements are similarity of velocity profile, $\overline{V}_z/\overline{V}_G$, turbulence intensity, $\sigma_V/\overline{V}_z$, and power spectral density of the longitudinal component, $n\bar{S}V(n)/(\sigma_V)^2$. These are discussed in detail in Chapter 3. An example of these characteristics is given in Fig. 2.22.

A variety of techniques have been used to achieve these minimum requirements. Basically the natural wind, for strong wind, neutral conditions is a rough wall turbulent boundary layer with some larger scales superimposed and hence is best modelled in the same way. The development of a rough wall boundary layer requires a long wind tunnel working section and a number of wind tunnels, sometimes called boundary layer wind tunnels, have been built for this purpose. Two types of wind tunnel and means of developing the natural wind models will be given as examples.

At Monash University the main wind tunnel is of the closed circuit type shown in Fig. 2.23. There are two working sections in

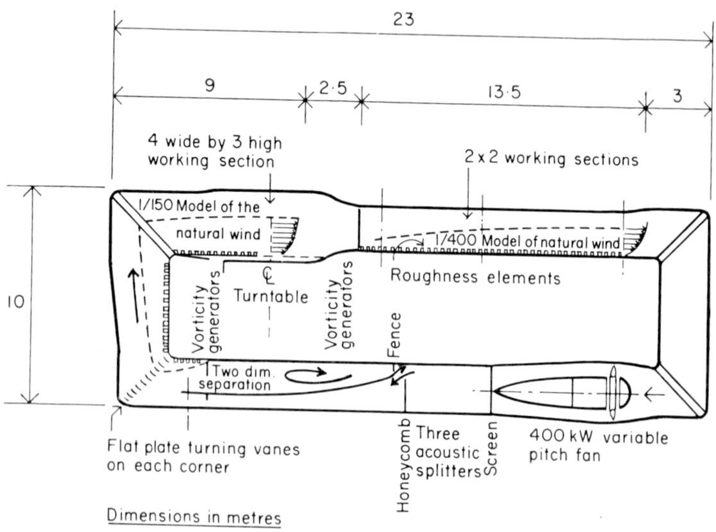

Fig. 2.23. Layout of the 400 kW wind tunnel at Monash University in which natural wind models are developed.

which natural wind models are developed. In the 4 × 3 m working section model scales of between 1/100 and 1/300 for a range of terrain conditions are developed by superimposing a large two-dimensional separation bubble on a rough wall boundary layer growing along three legs of the wind tunnel circuit. In the 2 × 2 m working section, following a contraction, a new shear layer is initiated with triangular

Fig. 2.24. Model buildings in the 4 × 3 m and 2 × 2 m working sections of the Monash University 400 kW wind tunnel set up to produce scale models of the natural wind in each section, for alternate use.

vorticity generators or a two-dimensional separation bubble which is superimposed on a rough wall boundary layer. Model scales of between 1/400 and 1/1200 have been developed in this working section for a variety of terrain conditions obtained by varying the size of the roughness elements and the initial conditions. A photograph of the two working sections of the Monash University wind tunnel is shown in Fig. 2.24.

The boundary layer tunnel at the University of Sydney is of the

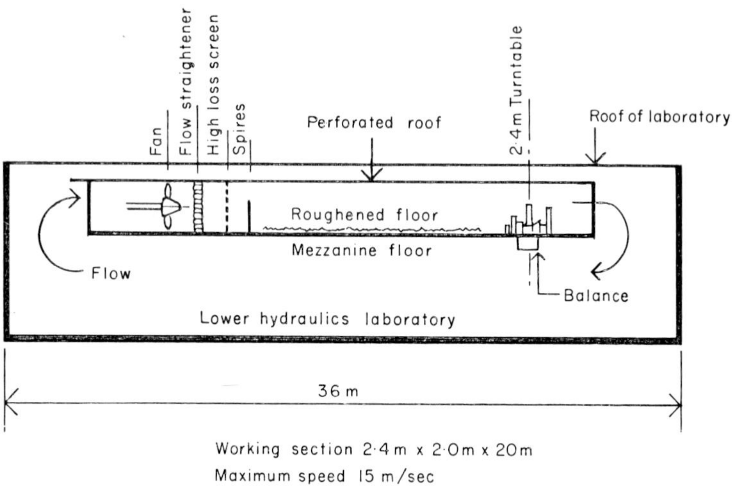

Fig. 2.25. Arrangement of the boundary layer wind tunnel at the Department of Civil Engineering, University of Sydney.

open circuit type (i.e. return air is circulated through the building), with approximately $2\frac{1}{2} \times 2$ m working sections as shown in Fig. 2.25. The natural wind model development used is similar to that described in the previous paragraph where the larger scale fluctuations from either vorticity generators or a trip board are superimposed on a rough wall boundary layer growing over roughness elements the size and spacing of which are varied to get different wind models.

2.11.3. Model Measurements

It is beyond the scope of this chapter on basic aerodynamics to go too deeply into the modelling and experimental techniques used to measure local and total wind loads on structures and the techniques used for studying flow fields and effluent disposal. In some cases these will be referred to in later chapters. However, some aspects of the model measurements are universal to this field of study and some mention of them should be made before leaving this section on wind tunnel testing.

The variables requiring measurement on the models are most commonly pressure, stress, force, moment, displacement (and higher derivatives), and the velocity (turbulence characteristics) of the model

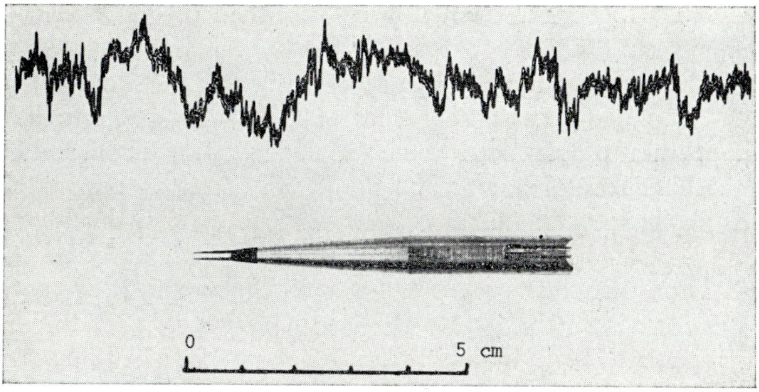

Fig. 2.26. 5 micron hot wire anemometer head and associated record of velocity.

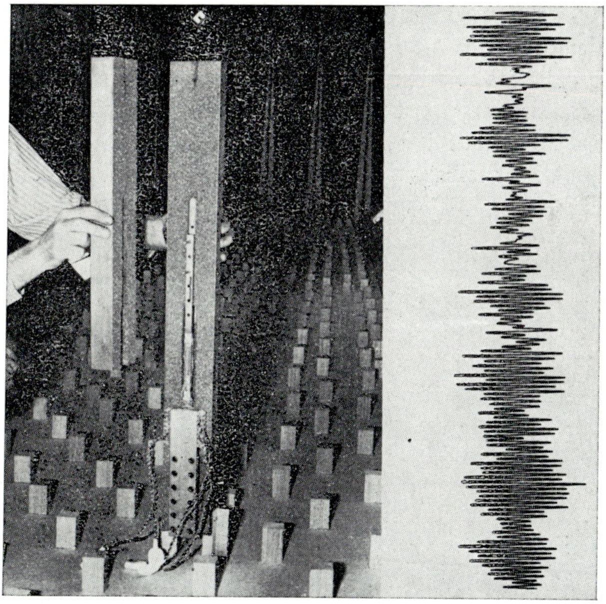

Fig. 2.27. Breakdown of a model building mounted on a strain gauge balance along with a trace of the cross-wind force output.

wind needs to be measured. In every case the variable is behaving randomly as a function of time. This means that it is necessary always to be able to measure mean and standard deviations as a minimum requirement. It follows then that transducers must be used to give a voltage signal proportional to the variable being measured. Because of the frequencies involved and the length of record required for analysis, the use of digital or hybrid facilities is virtually mandatory. Most wind tunnel facilities working on wind effects have access to digital facilities either on line or for analysing recorded data.

Velocity measurements are made using a hot wire or hot film anemometer. A typical velocity trace is shown in Fig. 2.26 along with the head of a hot wire anemometer used for making the measurement.

The hot wire anemometer is so called because it measures velocity fluctuations by detecting electrically the change in temperature of a heated wire caused by the change in fluid velocity.

Displacement, pressure and force measurements are made using a variety of transducer systems of which the strain gauge bridge system would be the most common. The strain gauge is a particularly useful device as it can be used to build up relatively simple special purpose balances to suit individual models. An example of a strain gauged cantilever support for a model building is shown in Fig. 2.27 with an output trace illustrating the random amplitude of response of this building oscillating at its fundamental natural frequency.

Chapter 3

Physical and Statistical Properties of the Wind

3.1. Major Storm Types

The energy source for the wind is the sun with the prime driving forces being associated with temperature differences and heat flow. Outside the tropics the majority of winds are associated with heat transfer from the equator to the poles through large scale low pressure systems or cyclones travelling from ~~east to west~~ WEST–EAST and having a circulatory flow clockwise in the Southern Hemisphere and counter-clockwise in the North. The temperature differences produce the basic north–south flow while the rotation of the earth generates the ~~east–west~~ WEST–EAST motion and determines the direction of circulation.

While the regular passage of cyclones dominates the wind climate of the temperate regions there are other storm types which, although not playing a major part in the wind climate as a whole, may contribute significantly to the extreme wind speeds. In particular there are many temperate regions in which the strongest winds are associated with comparatively small scale disturbances such as thunderstorms and tornadoes.

Both the isolated thunderstorm on a hot afternoon and the many storms which may be associated with an advancing cold front derive their energy from heat. Cool moist air carrying water or ice is convected upward to mix with the drier upper air; with evaporation and cooling the air mass loses its buoyancy and starts to sink. The resulting strong downdraught arrives at the ground and spreads forward producing strong wind gusts. The scale of such a thunderstorm cell is small although in the case of line squalls the front may extend for hundreds of kilometres. In contrast with cyclonic winds which may persist at strength for many hours the duration of the strong thunderstorm winds is seldom more than 1 h and is commonly much less. This contrast is illustrated in the two anemographs shown in Figs. 3.1 and 3.2.

The tornado, a small vertical funnel with winds of perhaps 100

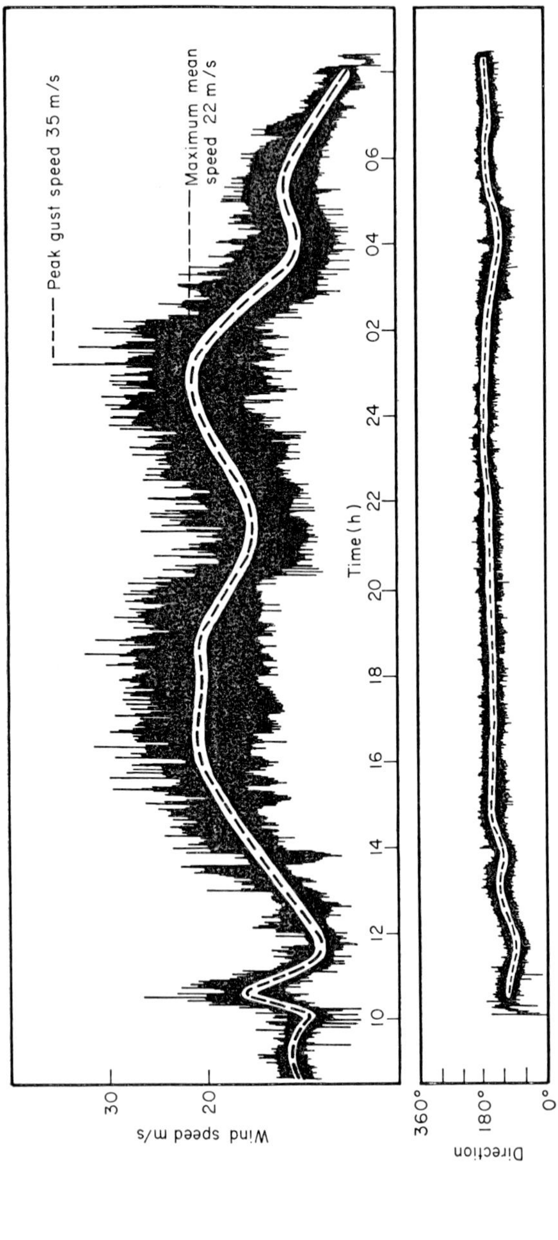

Fig. 3.1. The anemograph for an extra-tropical cyclone centred some hundreds of kilometres from the measuring point. Winds gusting in excess of 25 m/s persisted for more than 12 h.

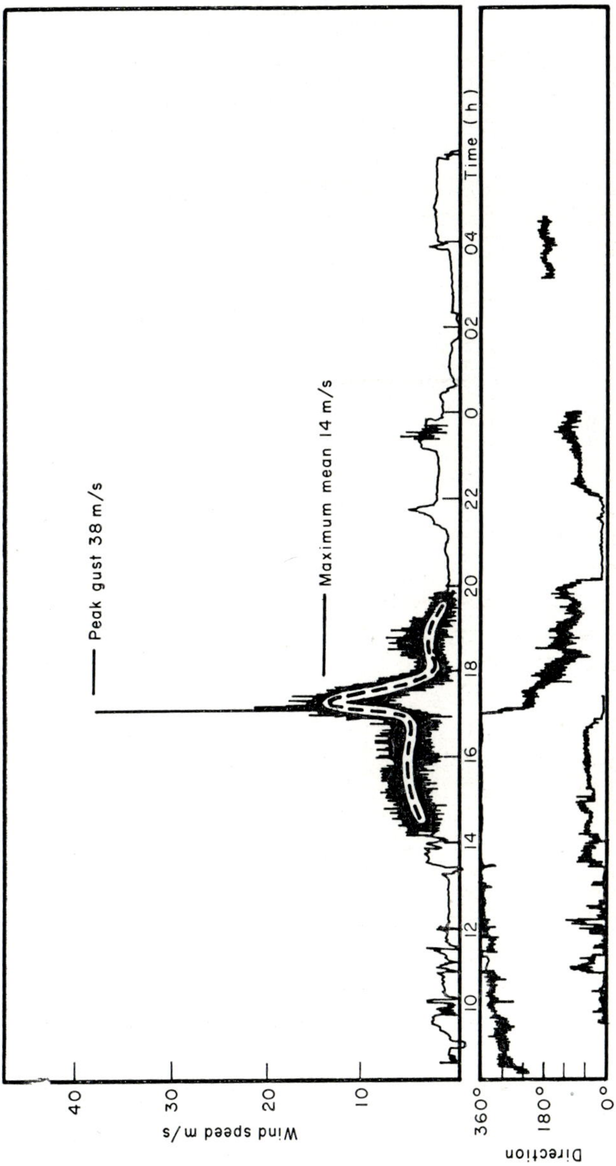

Fig. 3.2. An anemograph displaying the winds induced by a severe thunderstorm, the peak gust is isolated and appears to be almost unrelated to the speeds before and after. Winds gusting in excess of 10 m/s persisted for less than 30 min.

m/s or more revolving tightly about a core, is the most destructive of wind storms. Although widely distributed throughout the world, tornadoes occur most commonly in the continental plains of the U.S.A. and the U.S.S.R. Most of the quantitative data available concerns North American tornadoes which have a path length of typically [3.2] 5 km and a path width of 100–200 m These are average values, however, and path lengths in excess of 100 km and widths in excess of 500 m are not uncommon. While the area devastated by a single tornado may be small the weather conditions leading to their formation may extend over huge areas. The tornadic storms of 3–4 April 1974 illustrate this point; Fig. 3.3 [3.3] shows the recorded tornado paths during this period of some 24 h. There are 148 tornado tracks with a total path length in excess of 3000 km. These devastating storms resulted in some 330 fatalities and property damage in excess of $500 000 000·00.

In tropical and semi-tropical regions the wind climate is in general more placid than in the temperate zones but may be punctuated by infrequent storms of extreme intensity. Within perhaps 5° or 10° of the equator severe storms are rare but elsewhere in many coastal regions of the tropics the extreme winds are associated with tropical cyclones. Tropical cyclones, or hurricanes or typhoons as they are known in the Caribbean and China seas respectively, can develop and intensify only over a body of warm water. While tropical cyclones can and do move over land bodies or into cooler ocean regions they will, in both cases, degenerate and lose intensity. Over land the degeneration process is rapid and wind speeds of the same order as those existing over the sea will seldom penetrate more than 50 km or so inland.

While the size of tropical cyclones is far from constant an overall extent of 1000 km is typical, although winds of destructive intensity rarely extend for more than 100 km or so from the eye which itself might be 20 km or more in diameter. In tropical cyclones of exceptional strength the peak speeds may approach 100 m/s but speeds of perhaps 40 m/s are more typical.

The anemograph shown in Fig. 3.4 [3.1] was recorded during the passage of a tropical cyclone over the meteorological station. The circulatory nature of the flow resulted in a wind shift of nearly 180° as the eye passed almost centrally over the station. The comparatively

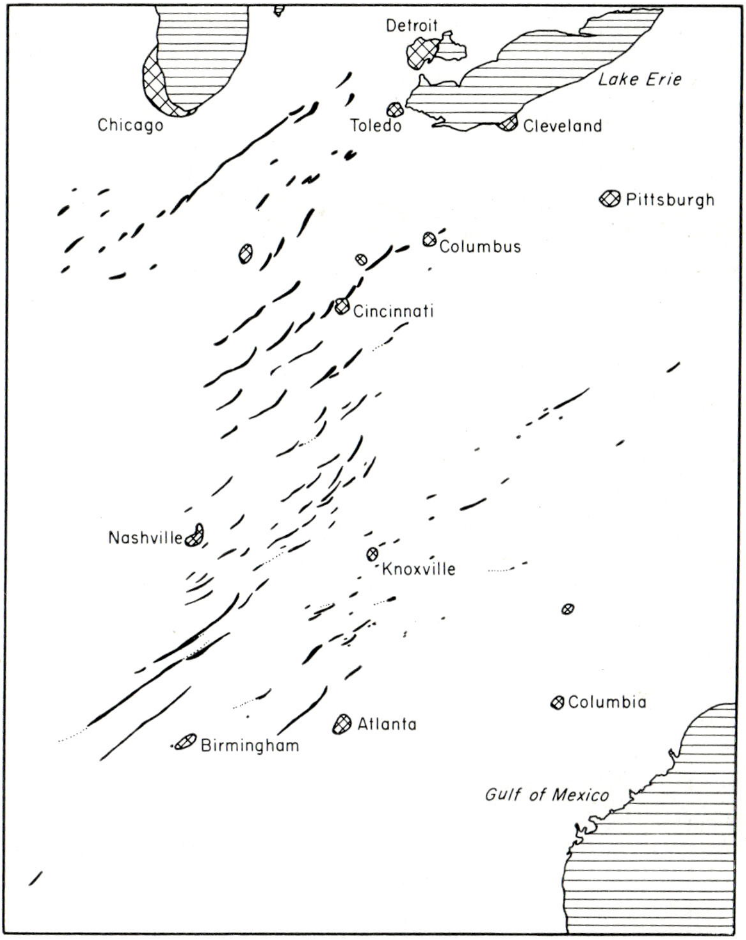

Fig. 3.3. Tornado paths 3–4 April 1974 (after Fujita [3.3]).

rapid reduction in speed following the passage of the eye was due to the reduction in speed with distance from the eye reinforced by the decay of the cyclone as it moved from the sea to a land mass.

The storm mechanisms introduced to this point are, between them, responsible for most wind storms of destructive intensity.

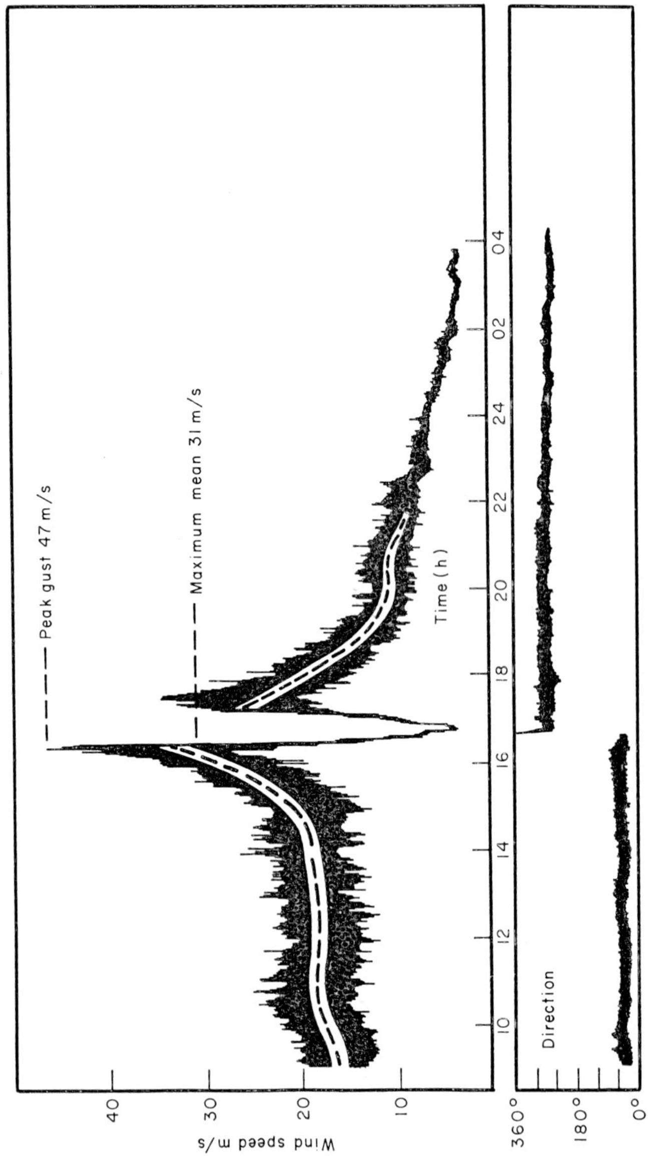

Fig. 3.4. Onslow anemogram of 4 March 1958. Example of a cyclone passing over a station.

There are of course other mechanisms and, in environmental problems rather than problems of structural integrity, these too are of importance. Perhaps the most significant of these mechanisms is the diurnal temperature variation which is responsible for a large part of the wind variation on comparatively still days and for the on-shore and off-shore breezes which can dominate the light wind climate of locations within a few kilometres of large bodies of water. The influence of these diurnal temperature variations is, however, barely significant in regard to winds of moderate or extreme strength.

3.2. Influence of Height and Terrain

The forces determining the flow pattern within the lower part of the atmosphere are:

(i) the forces of the pressure systems;
(ii) the so-called Coriolis forces which are associated with the rotation of the earth;
(iii) the frictional resistance provided by the rough boundary of the earth;
(iv) buoyancy forces associated with vertical temperature gradients;
(v) viscous forces.

In strong large scale systems it is only the first three types of force which play a significant role in determining the flow pattern and of these the frictional forces influence only the first 500 m or so of the atmosphere. At a height of about 500 m the wind speed and direction is a function of the form, intensity and geographic location of the pressure system. However, since it is the lower 500 m of the atmosphere which is of interest in the area of architectural aerodynamics it is clear the influence of ground roughness and surface friction must be examined.

3.2.1. The Boundary Layer

Before considering the details of the flow within the planetary boundary layer it is necessary to discuss the general nature of the flow within this region. It is clear from the broad traces of the anemographs of both the tropical and extra-tropical cyclones that the wind speed at the recorder was varying rapidly and widely in both magnitude and direction. It is also apparent from the general trend of the trace that the magnitude of the speed variations increases with the average speed defined by the trace centreline and that variations in this average speed occur comparatively slowly (although an exception to this behaviour is exhibited during the passage of the eye). The pattern exhibited in the two anemographs is typical of that observed for large scale storms and permits a quantitative description by way of two quantities, the mean speed and the gustiness. As in Chapter 2 the mean may be defined in mathematical terms as:

$$\overline{V} = \frac{1}{T} \int_{t_0 - T/2}^{t_0 + T/2} V_t \, dt \qquad (3.1)$$

where V_t = the instantaneous wind speed at time t, $\overline{V}$ = the mean speed averaged over some suitable time interval, T, centred on t_0.

It is convenient and mathematically acceptable to define $\overline{V}$ over time periods from about 10 min to 1 h. The gustiness or turbulence may be defined as:

$$\sigma_V{}^2 = \frac{1}{T} \int_{t_0 - T/2}^{t_0 + T/2} (V_t - \overline{V})^2 \, dt \qquad (3.2)$$

where σ_V = the standard deviation or root-mean-square value of deviations of the instantaneous speed from the mean.

The relative gustiness or the intensity of turbulence will be denoted by I and is defined as:

$$I_V = \frac{\sigma_V}{\overline{V}} \qquad (3.3)$$

In Fig. 3.5 is an expanded plot of a section of an anemograph such as that in Fig. 3.1. The values of $\overline{V}$ and σ_V are plotted in Fig. 3.5 but it is clear that these two quantities do not define completely the wind speed variations. The variations are random in character and cannot be defined in a deterministic fashion, it is only possible to define them in a statistical sense. The mathematical tool employed to describe these variations in speed is the probability function, P_V, or its derivative, the probability density function, p_V. The function P_V is defined such that: P_V = the probability that the wind speed is less than V or, somewhat loosely, the fraction of time during which the speed is less than V.

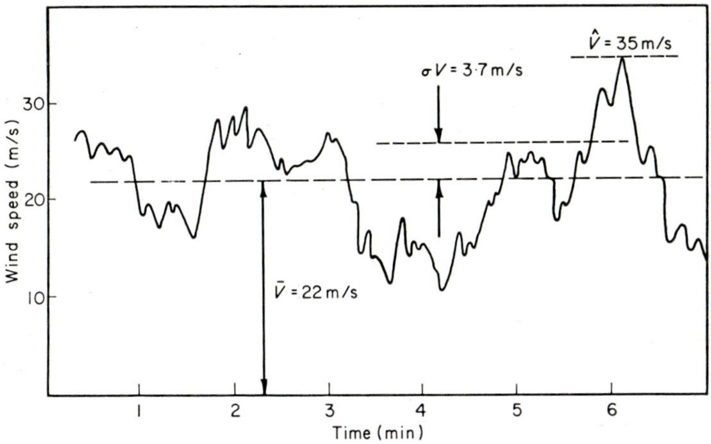

Fig. 3.5. Expanded plot of a section of the anemograph of Fig. 3.1.

The probability density function is such that: $p_V\,dV$ = probability that the speed at some time t lies in the interval V to $V+dV$. It follows then that the two describing functions are related as follows:

$$p_V = \frac{dP_V}{dV} \tag{3.4}$$

Measurements have shown that the distribution of wind speed in the atmosphere follows closely the 'Normal' or Gaussian distribution which is of the form:

$$p_V = \frac{1}{(2\pi)^{1/2}\sigma_V} \exp\left[-\left(\frac{V-\bar{V}}{\sigma_V}\right)^2\right] \tag{3.5}$$

With the knowledge that the probability distribution is Gaussian it is clear that the mean $\bar{V}$ and the standard deviation, σ_V, are sufficient to describe statistically the variations in the wind speed over some time interval T.

The probability distribution of the trace in Fig. 3.5 is plotted in Fig. 3.6 and from this we can deduce that for a mean speed of 22 m/s

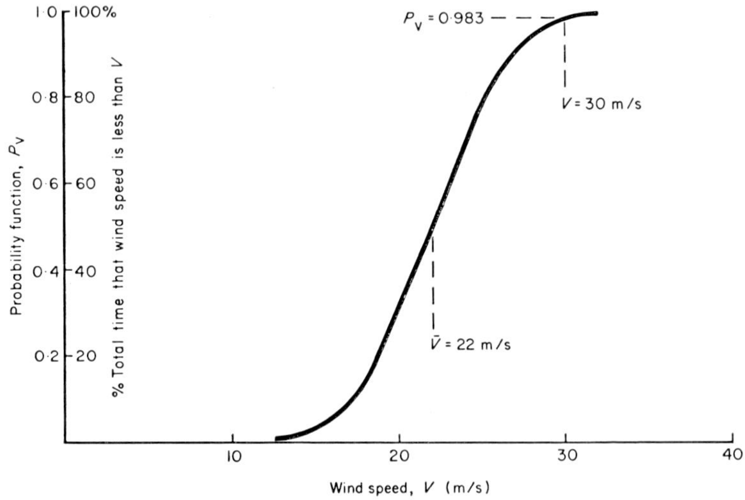

Fig. 3.6. Probability distribution of wind speed shown in Fig. 3.5.

and a standard deviation of 3·7 m/s a speed of 30 m/s will be exceeded for a total of about 1 min/h (i.e. $P_V = 1-1/60 = 0.983$). It is also possible to determine the expected maximum speed in a given period but such a development is beyond the scope of this treatment. It suffices to state that for periods of from 10 min to 1 h the peak gust as recorded by most meteorological anemometers will exceed the mean by about 3·5 σ_V.

3.2.2. *Mean Speeds*

Having defined the parameters $\overline{V}$ and σ_V the structure of the flow within the boundary layer may be described in a quantitative sense. At some height, z_G, the mean wind $\overline{V}_G$ is independent of the ground effects and is dependent only upon the pressure field or gradient and its location. The speed $\overline{V}_G$ is known as the gradient speed and the height at which it is attained, the gradient height z_G. As might be expected, the rougher the terrain the greater the value of z_G and the more pronounced the effects of boundary shear. The gradient height

Table 3.1

Values of gradient height, z_G*, power law exponents* α*,* β *and the roughness length* z_0

Terrain category	Terrain description	Gradient height, z_G (m)	Roughness length, z_0 (m)	Mean speed exponent α	Gust speed exponent β
1	Open sea, ice, tundra, desert	250	0·001	0·11	0·07
2	Open country with low scrub or scattered trees	300	0·03	0·15	0·09
3	Suburban areas, small towns, well wooded areas	400	0·3	0·25	0·14
4	Numerous tall buildings, city centres, well developed industrial areas	500	3	0·36	0·20

varies from about 250 m over open sea, ice fields, tundra and other 'smooth' terrains to perhaps 400–500 m over an extremely rough city terrain. The variation of $\overline{V}_z$, the mean speed at some height z, from the ground to gradient height can be approximated by a number of mathematical expressions but it is the 'power law' and 'logarithmic law' which are in common use. In the 'power law' representation,

$$\frac{\overline{V}_z}{\overline{V}_{zG}} = \left(\frac{z}{z_G}\right)^{\alpha} \tag{3.6}$$

in which z_G and α are functions of ground roughness. Typical values of z_G and α are given in Table 3.1 and the corresponding mean wind speed profiles are plotted in Fig. 3.7.

The 'logarithmic law' is of the form:

$$\bar{V}_z = \bar{V}_1 \log_e \left(\frac{z}{z_0}\right) \Big/ \log_e \left(\frac{z_1}{z_0}\right) \quad (3.7)$$

in which $\bar{V}_1$ is the mean speed at some reference height z_1 and z_0 is a roughness length which has a value equal to about 5–10% of the

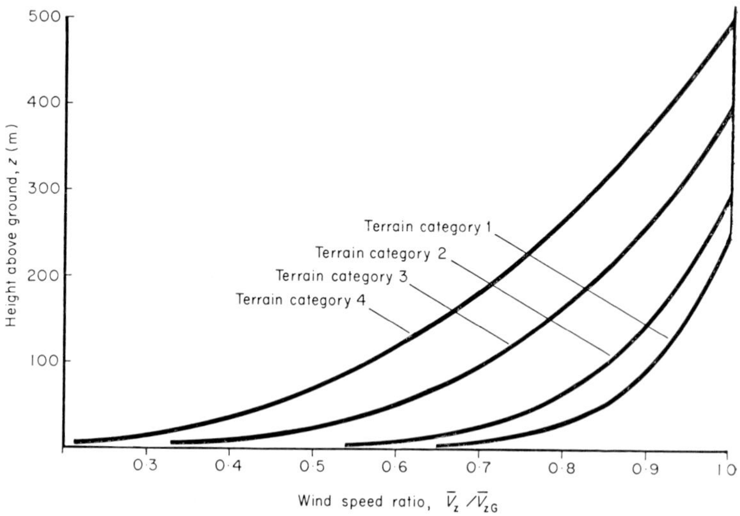

Fig. 3.7. Mean wind speed profiles.

average height of the terrain roughness elements (e.g. houses, trees, etc.). A somewhat improved form of these 'laws' can be obtained if the heights are measured not from the true ground but a false ground which varies with the height of the roughness elements of the terrain. This correction can be significant in densely developed areas where the effective ground plane tends to be nearer the average roof height rather than at true ground level.

It is clear from the profiles plotted in Fig. 3.7 that the mean speed near the ground is heavily dependent on terrain roughness. This influence is illustrated by the comparison between Figs. 3.1 and 3.8 which are anemographs for the same storm recorded at two different stations separated by about 15 km. The trace in Fig. 3.1 was obtained

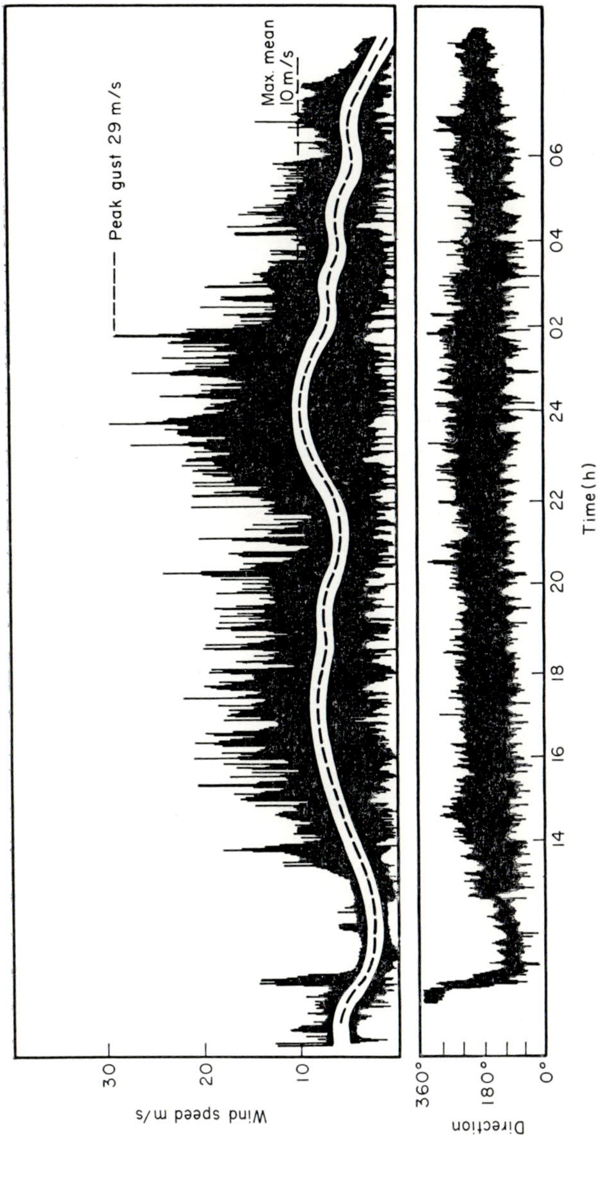

Fig. 3.8. Anemograph for a strong extra-tropical cyclone; recorded immediately downwind of a city centre.

from an anemometer in flat open country within 1 km of open water while the trace in Fig. 3.8 was obtained from an anemometer at the edge of and downwind from a large city (Sydney) with numerous buildings over 100 m in height. The ratio of the maximum mean speeds is about 2·2:1 although the anemometer heights of 10 m and 18 m are comparable. This ratio is in good agreement with the profiles plotted in Fig. 3.7 which suggest a ratio of 0·70 $\bar{V}_G$:0·31 $\bar{V}_G$ or 2·3:1. The ratio of mean wind pressures at the two sites is more dramatic since this varies with the square of wind speed and has a value of almost 5:1.

3.2.3. Turbulence Levels

In a given storm with a mean gradient $\bar{V}_G$ the magnitude of the turbulent fluctuations, σ_V, does not exhibit the changes with height and terrain shown by $\bar{V}$. As a first approximation σ_V is independent of both height and roughness and has a value of about 0·10 $\bar{V}_G$. Closer examination would indicate that σ_V tends to increase slightly with terrain roughness and decrease slightly with height but these variations are not marked. If the range of the wind speed fluctuations rather than the standard deviation is accepted as a measure of turbulence then Figs. 3.1 and 3.8 give an indication of the turbulence in the same storm but at two locations with extreme differences in roughness. From Fig. 3.1 it can be seen that in a smooth terrain the range near the peak of the storm was about 25 m/s while at the very rough city location the range was only marginally higher at 29 m/s.

The intensity of turbulence ($i = \sigma_V/\bar{V}$) does differ markedly, primarily because of the change in $\bar{V}$. This is illustrated by the ratio of the range to the mean which varies from 1·1 in the smooth terrain to 2·9 in the rough city terrain.

The turbulent intensity is reflected not only in the range to mean ratio but also in the variations in direction or equivalently the variations in the ratio $\hat{V}_c/\bar{V}$ where $\hat{V}_c$ is the peak value of the cross-wind component of turbulence and $\bar{V}$ the mean flow velocity measured in the direction of the mean flow. In the smooth terrain the direction fluctuates through about ±20° typically and perhaps ±40° maximum. In the rough city terrain the fluctuation is about ±60° typically and the peak fluctuations correspond to a complete reversal of the flow direction.

3.2.4. *Gust Speeds*

If the power law variation of the mean speed is adopted together with the simple concept of $\sigma_V = 0{\cdot}10\ V_G$ the peak or gust speed at some height z can be expressed approximately as:

$$\hat{V}_z = \bar{V}_z + 3.5\,\sigma_V$$

$$= \bar{V}_z + 0{\cdot}35\,\bar{V}_G$$

or

$$\frac{\hat{V}_z}{\hat{V}_{zG}} = \frac{1}{1{\cdot}35}\left\{\left(\frac{z}{z_G}\right)^{\alpha} + 0{\cdot}35\right\} \qquad (3.8)$$

An alternative representation which is sometimes employed is a power-law of the form:

$$\frac{\hat{V}_z}{\hat{V}_G} = \left(\frac{z}{z_G}\right)^{\beta} \qquad (3.9)$$

in which $\hat{V}_z$ = peak or gust speed at height z; $\hat{V}_G$ = peak or gust speed at the gradient height z_G; β = an exponent which varies with the terrain roughness and has a value equal to about $0{\cdot}6\,\alpha$; and

$$\hat{V}_{zG} \simeq 1.35\,\bar{V}_{zG}. \qquad (3.9a)$$

It is obvious that eqns. (3.8) and (3.9) cannot be identical but, as is shown in Fig. 3.9, the differences are small if β is suitable. Values of

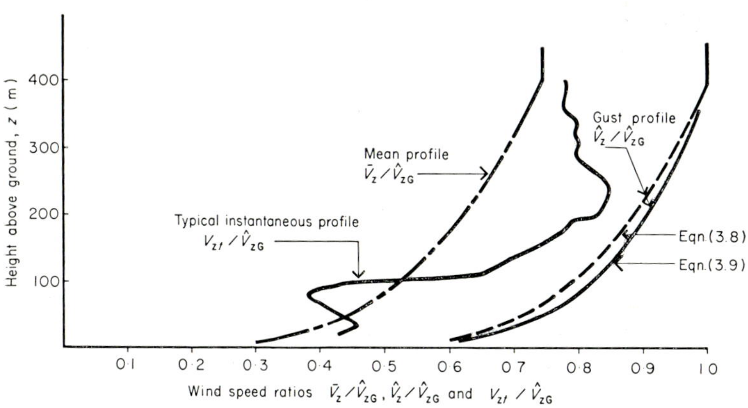

Fig. 3.9. Mean and gust speed profiles for Category 3 terrain.

β for the four terrain categories are given in Table 3.1. The profiles as defined by eqns. (3.8) and (3.9) are not real profiles in the sense that they exist at any given instant. They are profiles of the 'envelope' of peak gust speeds rather than the profile at a particular time. At any given time, t, the instantaneous profile might be as indicated in Fig. 3.9 as $V_{z,t}/\hat{V}_{zG}$.

The characteristics of the wind speed profiles discussed up to this point are illustrated in Fig. 3.10 [3.4] which is a record from a tall

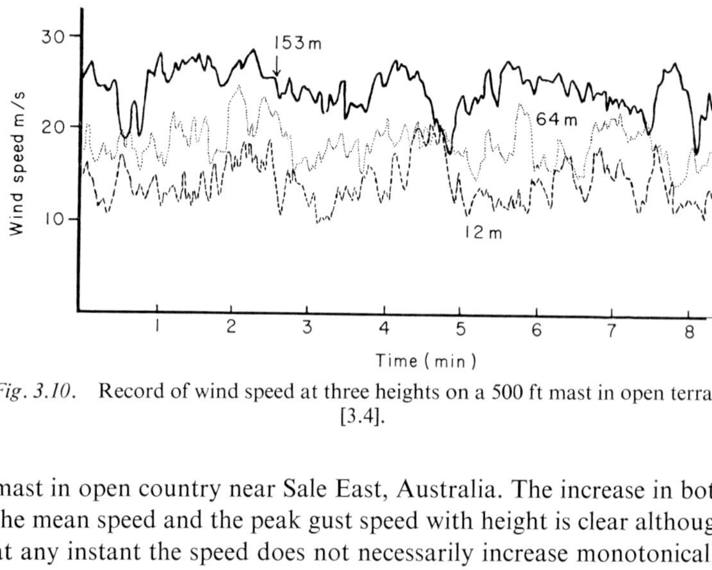

Fig. 3.10. Record of wind speed at three heights on a 500 ft mast in open terrain [3.4].

mast in open country near Sale East, Australia. The increase in both the mean speed and the peak gust speed with height is clear although at any instant the speed does not necessarily increase monotonically with height. There are a number of instances where the traces cross thus indicating a higher speed at the lower measuring point.

3.3. Wind Speed Spectra

The probability distribution introduced in this chapter permits a description of the magnitude of a random or stochastic process such as the variation of wind speed with time but it adds nothing to our knowledge of how quickly or slowly this process might vary in time. The probability distribution on the process illustrated in Fig. 3.1 or

Fig. 3.6 is quite independent of the time scale. In order to define the process in the time rather than the amplitude domain, another describing function is required. The function which serves this purpose is the spectral density function or spectrum of the process. Spectrum of longitudinal turbulence represents the distribution of turbulence energy over a frequency range. It can be defined by the spectrum, $S(n)$, such that

$$\sigma_V{}^2 = \int_0^\infty S_V(n)\,\mathrm{d}n$$

or $S_V(n).\Delta n$ = contribution to the variance ($\sigma_V{}^2$) from frequency components in the range n to $n+\Delta n$ as shown in Fig. 3.11; where $\sigma_V{}^2$ is the mean square value of the fluctuating component of the wind speed V; and $S_V(n)$ is the power spectral density of the fluctuating component of the wind speed V.

3.3.1. Long Term Spectra

Figure 3.11 shows such a spectrum determined from 55 years of wind record from an anemometer in Sydney, Australia [3.5]. The frequency range of this spectrum extends from approximately $\frac{1}{10}$ cycle/year to about 3 c/s. In this instance the spectrum is of variations in speed about the long term average rather than an average over about 10 min to 1 h.

There are four major peaks in the spectrum of Fig. 3.11 and each of these can be associated with a physical phenomenon. The peak at 10^{-4} c/h or 1 cycle/year is created by the seasonal variations in wind speed. The multiple peak near 7×10^{-3} c/h or one cycle in 6 days is created by the passage of large scale pressure or weather systems which, on their eastward path, pass through Sydney once every six days on average. The peak at 0·04 c/h or 1 cycle/day is generated by the diurnal variations in wind speed caused by the daily heating/cooling cycle. The final broad peak which reaches a maximum at about 30 c/h or 1 cycle in 2 min is related to the nature of atmospheric turbulence rather than any slow variations in the general wind level.

Long term spectra measured at other points have shown similar characteristics although there are differences in detail. In inland situations or at points well above ground the diurnal peak will not be as pronounced as in Fig. 3.11 for which the anemometer is located near a land/sea boundary and about 60 m above sea level. Van der Hoven's [3.6] measurements at Brookhaven (New York) indicate a peak at 1 cycle in 4 days rather than 6 days as for Sydney. A notable feature of the spectrum in Fig. 3.11 and spectra measured by Van

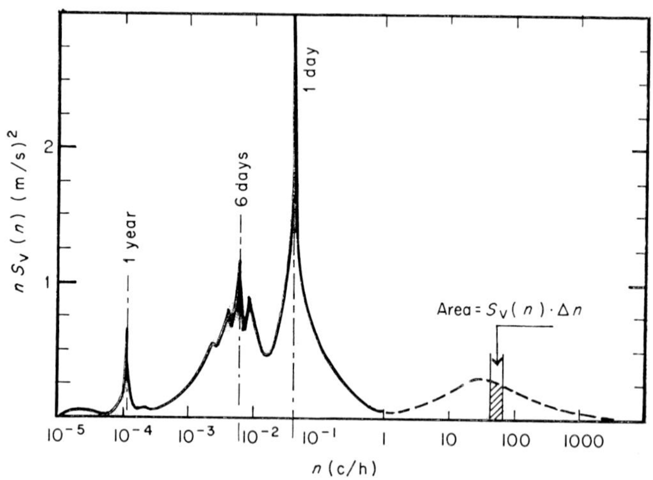

Fig. 3.11. Long-term spectrum of wind speed (Sydney, Australia).

der Hoven and others is the gap at about 1 c/h which enables a convenient distinction to be made between the mean wind speed $\overline{V}$, averaged over say 10 min to 1 or 2 h, and the superimposed gusts or atmospheric turbulence.

3.3.2. *Turbulence Spectra*

The separation of the mean wind and the gusts permits a somewhat different definition of the high frequency range of the total wind spectrum. The intensity of gusting, σ_V, at a given location is proportional to the mean speed $\overline{V}$ while the time taken for a gust of given size to pass an observing point will be inversely proportional to $\overline{V}$. These simple observations suggest that if the spectrum $nS(n)$ was plotted as a non-dimensional quantity, say $nS(n)/\overline{V}^2$, and the frequency was replaced by a dimensionless parameter $nL/\overline{V}$, where

L is a characteristic length, perhaps the depth of the boundary layer, then a universal spectrum might result.

Studies by Davenport [3.7] of a number of gust or turbulence spectra indicated that the spectrum could be expressed as:

$$\frac{nS(n)}{k\bar{V}_{10}^{2}} = \frac{4{\cdot}0\left(\dfrac{nL}{\bar{V}_{10}}\right)^{2}}{\left\{1+\left(\dfrac{nL}{\bar{V}_{10}}\right)^{2}\right\}^{4/3}} \tag{3.10}$$

where $L \simeq 1200$ m; k = a drag coefficient which is dependent on the terrain roughness and has values of about 0·001, 0·005, 0·015 and 0·04 for terrain categories 1–4 respectively. $\bar{V}_{10}$ = mean speed at height of 10 m.

The spectrum defined by eqn. (3.10) is plotted in Fig. 3.12 together with a number of field observations. More recently it has been suggested that the characteristic length L and the coefficient k should be related to the height of the observation point and the roughness of the terrain. While this could well be correct it is not intended to use the spectrum in a strict quantitative sense but merely to employ it to define some significant features and to this end eqn. (3.10) is quite adequate.

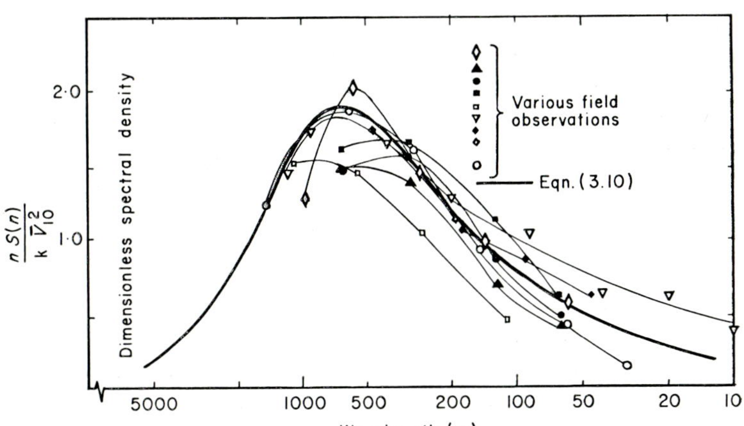

Fig. 3.12. Spectrum of turbulence in strong winds (after Davenport [3.7]).

The spectrum indicates that the energy of atmospheric turbulence is associated primarily with wavelengths of the order of 600 m and that the available energy decreases monotonically with decreasing wavelength. There is very little energy at wavelengths less than about 10 m. At wind speeds of interest in terms of wind loading, say 30 m/s, a wavelength of 600 m implies a frequency of about 3 c/min and a wavelength of 10 m implies a frequency of 3 Hz.

3.3.3. Gust Size

The concept of a wavelength, $\lambda = \overline{V}/n$, permits the association of a length with a particular frequency component. The definition of a gust size however is somewhat more complex. If it were possible to isolate gusts of a given wavelength it would be found that the vertical and lateral extent varied from gust to gust in a random fashion and it is possible to talk only of average sizes. On average the lateral extent is about $\lambda/8$ and the vertical extent $\lambda/5$. In the case of the major or energy containing gusts with a wavelength of from 500 to 1000 m the average across-wind dimensions are of the order of 100–200 m. The significance of gust size in the evaluation of wind loads is discussed in Chapter 4; at this stage it suffices to point out that because of size limitations not all gusts will be fully effective in producing load on large structures.

3.4. Topographic Effects

The discussion of the variation of wind speed with height and terrain in this chapter has been directed toward situations where there are no marked topographic features such as ridges, hills and valleys. While the literature, and indeed some codes, contain warnings as to the extreme speeds that can be induced by topographic features very little quantitative data is available.

The Australian Wind Code [3.8] employs a very simple approach which, although not altogether adequate, does make a reasonable allowance for the increase in wind speed over hills and ridges. The method adopted in this code and others is to use an artificial datum from which the height is measured. The design wind speed is then chosen as if the terrain were level but with a height measured from

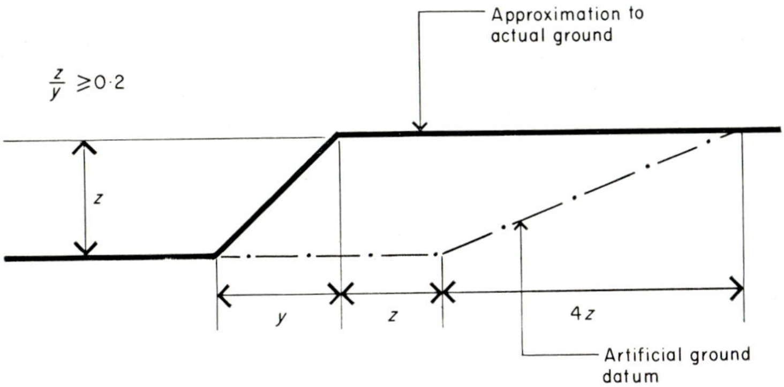

Fig. 3.13. Definition of artificial ground datum near an escarpment. (Australian Wind Code.)

the datum rather than the local ground level. The definition of this artificial datum is illustrated in Fig. 3.13.

Figure 3.14 shows some wind tunnel data obtained by De Bray [3.9]. The wind speed ratios plotted in this figure are *mean* speed ratios where:

$\bar{V}_1$ = mean speed of approach flow at some height z;

$\bar{V}_2$ = mean speed of flow at the section indicated and at a height z above local groud level.

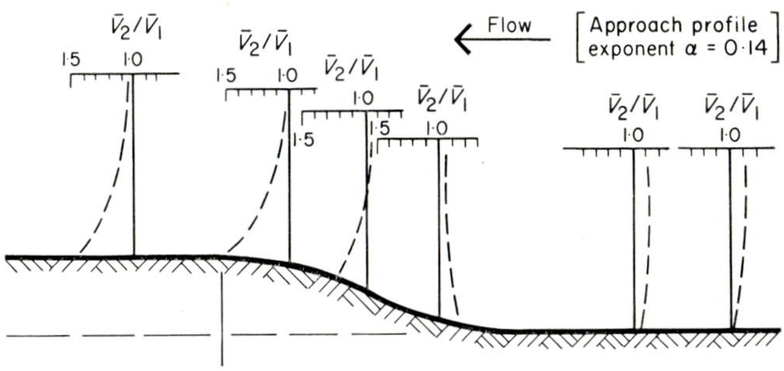

Fig. 3.14. Wind speed ratios near an escarpment [3.9]. $\bar{V}_2$ = measured speed at height z above ground; $\bar{V}_1$ = measured speed in approach flow at height z.

The magnification near the top of the ridge is almost 1·5 although if attention was directed towards the peak gust speeds the amplification would not be as great. In circumstances such as those illustrated the turbulence σV would tend to be unchanged and hence the gust speed ($\hat{V} = \bar{V} + 3{\cdot}5\,\sigma V$) would increase by perhaps 30% rather than 50%.

Further wind tunnel tests are presented in Fig. 3.15. In this case

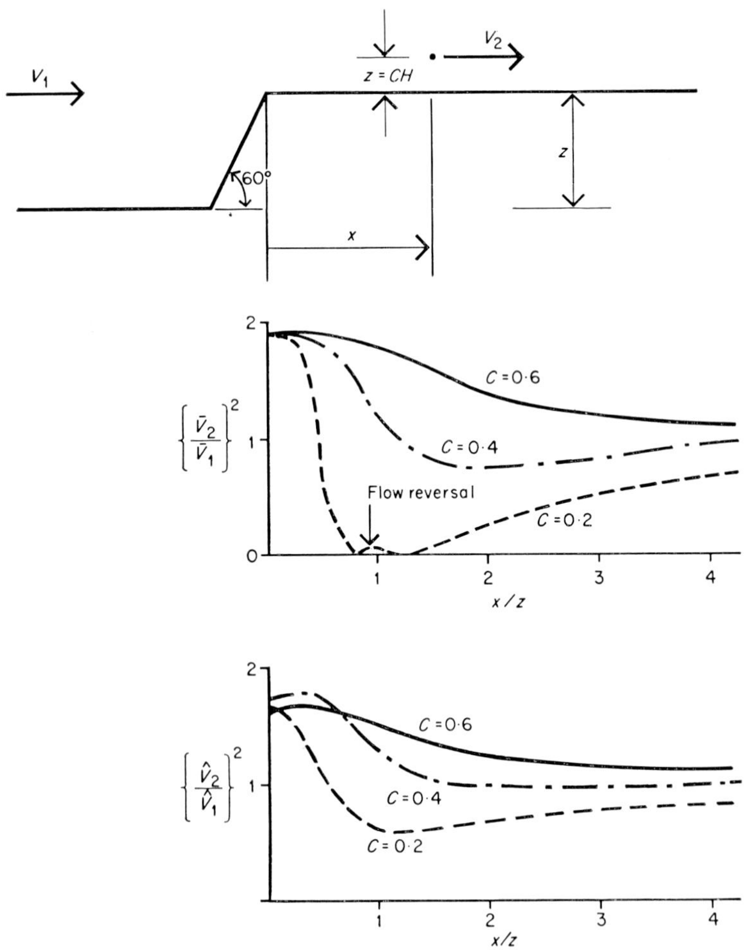

Fig. 3.15. Mean and gust speeds near a sharp escarpment.

the 60° escarpment is sharp edged and the flow separates from the edge of the cliff and re-attaches some distance back from the edge. Both mean and gust speed ratios are plotted in Fig. 3.15. Within the separation region the mean flow is actually reversed although the peak gusts are still in the direction of the main flow. Very near the cliff edge both the mean and gust speeds are substantially greater than the corresponding values measured in the approach flow at cliff height. At a distance of 4 cliff heights from the edge the wind speeds near local ground are very nearly equal to those in the undisturbed approach flow at cliff height.

3.5. Wind Climate

In Sections 3.1–3.4 discussion has centred on the structure of the wind, its gustiness and the influences of terrain and topography on wind strength. This section is concerned with the description of the absolute wind strength at a particular site or the wind climate of that site.

3.5.1. Probability Distributions

The simplest statement of wind climate is the probability distribution of wind speed without regard to direction. For this purpose the so-called Weibull distribution has been found to fit a wide collection of wind data. The Weibull distribution is of the form:

$$P(V) = 1-\exp\left[-\left(\frac{V}{C}\right)^k\right] \tag{3.11}$$

or

$$P(>V) = \text{probability of the wind speed exceeding } V$$

$$=\exp\left[-\left(\frac{V}{C}\right)^k\right] \tag{3.12}$$

With a double logarithmic transform this distribution may be written as:

$$\log_e(-\log_e P(>V)) = k\log_e V - k\log_e C$$

and hence a plot of $\log_e V$ against $\log_e(-\log_e P(>V))$ should yield a straight line if the Weibull distribution is a suitable choice for a particular data set.

Figure 3.16 shows nine probability distributions plotted as $\log_e V$ against $\log_e(-\log_e P(>V))$ and it can be seen that a straight line provides an adequate fit to each data set. From plots of this type the

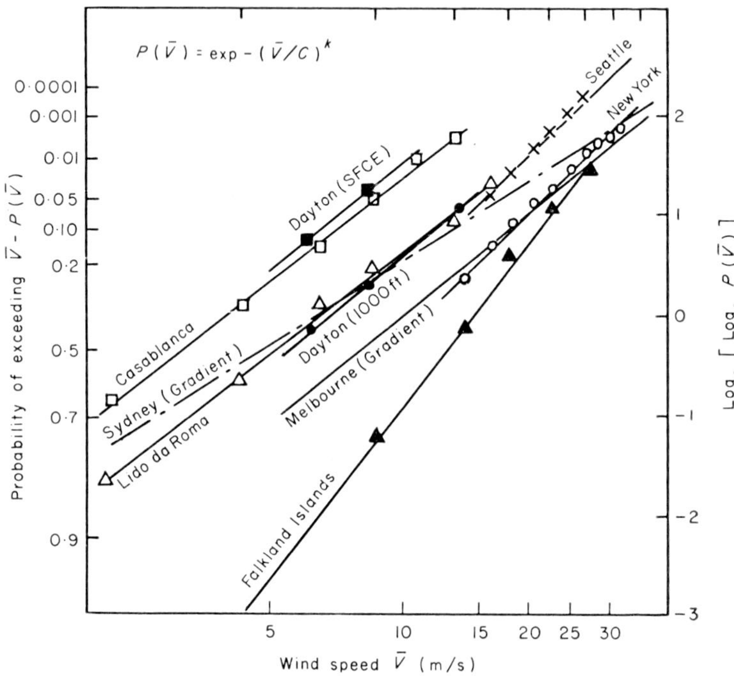

	C (m/s)	k
Dayton (Ohio) surface	3·9	1·7
Dayton (Ohio) (300 m)	7·1	1·7
Casablanca	4·2	1·5
Falkland Islands	14·8	2·5
Lido da Roma	6·8	1·5
New York (F.J.K. airport—500 m)	11·4	1·87
Seattle airport (300 m)	8·5	1·9
Sydney (gradient)	7·8	1·39
Melbourne (gradient)	10·5	1·64

Fig. 3.16. Weibull distribution of wind speed.

Weibull parameters C and k can be determined. The parameter C is in the nature of a characteristic speed for the site.

The Weibull distribution can also be used to describe the distribution of wind speed and direction. Climatological data can be gathered and then in each of say sixteen directions a Weibull function fitted. This directional distribution can be expressed as:

$$P(>V, \theta) = A(\theta) \exp\left[-\left(\frac{V}{C(\theta)}\right)^{k(\theta)}\right] \tag{3.13}$$

The climatological data presented in Table 3.2 was collected from meteorological balloon observations in Melbourne, Australia. These data were fitted by Weibull distributions to yield the directional distribution shown in Fig. 3.17 and also the Weibull distribution of wind speed which is included in Fig. 3.16. Distributions such as that shown in Fig. 3.17 are essential in studies of ground level wind environment where wind tunnel studies are employed to relate ground

Table 3.2

Distribution of wind speed and direction (800 m, Melbourne, Australia). Total no. of observations: 20 580 over 20 years)

	Percentage of observations								
Direction	0·5–5 (m/s)	5–10 (m/s)	10–15 (m/s)	15–20 (m/s)	20–25 (m/s)	25–30 (m/s)	30–35 (m/s)	35–40 (m/s)	Total
NNE	0·957	0·996	0·428	0·141	0·058	0·024	—	—	2·60
NE	0·792	0·326	0·049	0·024	—	—	—	—	1·19
ENE	0·850	0·287	0·015	—	—	—	—	—	1·15
E	1·696	0·685	0·136	—	—	—	—	—	2·52
ESE	1·283	1·093	0·374	0·073	0·010	—	—	—	2·78
SE	1·667	2·167	0·437	0·058	0·010	—	—	—	4·34
SSE	1·822	2·843	0·496	0·029	0·005	—	0·005	—	5·20
S	2·658	5·510	1·433	0·214	0·019	—	—	—	9·84
SSW	1·433	2·852	1·268	0·301	0·010	—	—	—	5·86
SW	1·574	3·324	1·798	0·525	0·058	—	—	—	7·28
WSW	1·351	3·285	2·420	1·040	0·185	0·024	0·019	—	8·32
W	1·919	3·984	3·290	1·501	0·345	0·034	0·010	0·005	11·09
WNW	1·370	2·736	2·138	0·972	0·199	0·049	0·005	—	7·46
NW	1·370	2·648	2·182	1·239	0·500	0·121	0·023	0·005	8·10
NNW	1·385	3·017	2·609	1·885	0·850	0·394	0·078	0·015	10·24
N	1·963	3·411	3·139	1·983	1·050	0·355	0·097	0·019	12·03

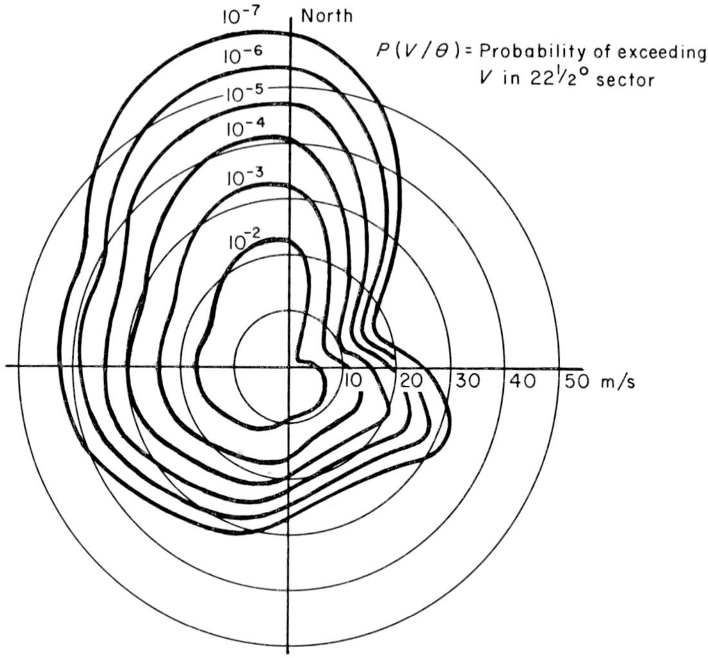

Fig. 3.17. A directional distribution of wind speed at 800 m.

level wind speeds to a reference speed such as that at gradient height. The test results together with a distribution such as that in Fig. 3.17 can be integrated to yield the wind speed distribution or wind climate at ground.

3.5.2. Extreme Wind Speeds

The probability distributions discussed to this point are distributions of the whole population. A second and, in wind loading problems particularly, a more significant type of distribution is that dealing only with the larger values. If, for example, a data set comprising the largest wind speed recorded at some point in each of several years were obtained the distribution would be termed an extreme value distribution. If the parent or total population distribution from which these extremes were drawn was of an exponential type such as the Weibull then it can be shown [3.10] that the extreme value distribution will tend to the form:

$$P(V_{\text{extreme}} > V) = 1 - e^{-e^{-a(V-U)}} \qquad (3.14)$$

where U is known as the modal wind speed and $1/a$ is wind speed known as the dispersion. The method by which the parameters U and a are determined is best introduced by example. Table 3.3 shows a

Table 3.3

Annual maximum hourly wind speeds at Observatory Hill, Sydney

Rank (n)	Speed (m/s)	Year	$P = \frac{n}{N+1}$	$-\log_e (\log_e P)$
1	12·0	1954	0·050	−1·10
2	12·5	1958	0·100	−0·83
3	13·0	1950	0·15	−0·64
4	13·5	1953	0·20	−0·48
5	13·5	1959	0·25	−0·33
6	13·5	1962	0·30	−0·19
7	14·0	1963	0·35	−0·05
8	14·0	1955	0·40	+0·09
9	14·5	1960	0·45	+0·23
10	15·0	1969	0·50	+0·37
11	15·0	1966	0·55	+0·51
12	15·0	1956	0·60	+0·67
13	15·0	1952	0·65	+0·84
14	15·5	1951	0·70	+1·03
15	16·5	1961	0·75	+1·25
16	17·0	1957	0·80	+1·50
17	17·5	1966	0·85	+1·82
18	17·5	1967	0·90	+2·25
19	18·5	1968	0·95	+2·97

data set comprising nineteen observed annual maximum mean hourly speeds recorded in Sydney, Australia. The speeds are ranked in descending order from $n = 1$ to $n = N = 19$ and a probability level, $P = n/N+1$, calculated for each. The wind speed, V, is then plotted against $-\log_e(-\log_e(1-P))$. From the nature of eqn. (3.14) it is apparent that provided the distribution is of this form a straight line plot will result; the plot for the data of Table 3.3 is shown in

Fig. 3.18. The parameters U and $1/a$ can then be determined as shown in Fig. 3.18, and for this data set, $U = 14{\cdot}0$ m/s and $1/a = 1{\cdot}68$ m/s or,

$$1 - P(V_{\text{extreme}} > V) = e^{-e^{-(V-14\cdot0/1\cdot68)}} \tag{3.15}$$

In cases where the extreme values are annual maxima it is common to express the probability P as $P = 1/R$ where R is the so-called

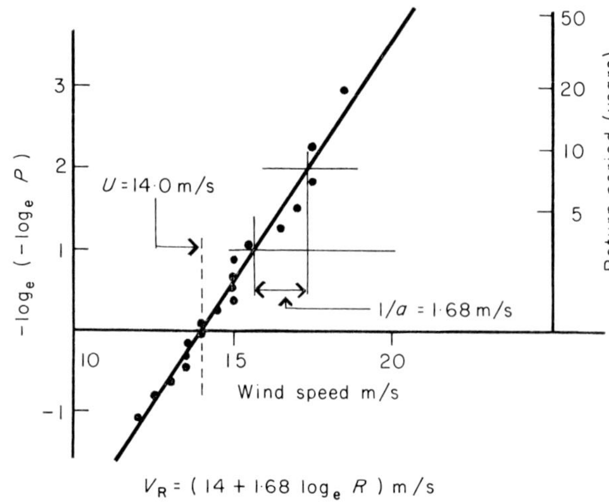

Fig. 3.18. Gumbel plot of annual maximum hourly mean speeds, Sydney.

return period in years. If for example $P = 1/20$ when $V = 19$ m/s then there is a probability of 1 in 20 that the annual maximum speed in any given year will exceed 19 m/s or, in a very long period of M years we could expect $M/20$ annual maxima to exceed 19 m/s. Another statement of this observation is that the annual maximum speed will exceed 19 m/s once every 20 years on average or that 19 m/s corresponds to a return period of 20 years. The actual intervals between years in which 19 m/s is exceeded will of course vary over a wide range and there is as much chance of the speed being exceeded in two successive years as there is in two full years separated by any other specific interval. If the concept of a return period R is introduced into eqn. (3.14) it follows that

$$1-\frac{1}{R} = e^{-e^{-a(V-U)}}$$

or

$$\log_e\left(1-\frac{1}{R}\right) = -e^{-a(V-U)}$$

and if R is large (>5, say) $\log_e(1-1/R)$ is very nearly equal to $-1/R$ and hence:

$$\frac{1}{R} \simeq e^{-a(V-U)}$$

or

$$-\log_e R \simeq -a(V-U)$$

or

$$V \simeq U+\frac{1}{a}\log_e R \tag{3.16}$$

For the data set in Table 3.3. it follows that:

$$V(R) = 14{\cdot}0+1{\cdot}68 \log_e R \text{ m/s}$$

and from this relationship the wind speed $V(R)$ corresponding to any required return period may be computed.

Chapter 4

Wind Loads and Structural Response

4.1. Introduction

In an introductory text in Architectural Aerodynamics it is neither feasible nor desirable to cover the field of wind loading in sufficient depth to bring the reader to the point of solving particular problems in detail. Even in a more specialised text such as Reference 4.1 it is not practical to cover such specialised subjects as the response of slender towers and bridges to wind forces. It is the aim of this chapter to develop an understanding of the nature of wind loads and their effects and to draw attention to situations which might be critical in regard to wind load.

4.2. Dynamics of Structures

It is clear from the description of the wind given in Chapter 3 and from the description of the flow field of bluff bodies given in Chapter 2 that wind loading is far from steady or static. Indeed, the most apparent feature of the wind and therefore wind pressure and load is its dynamic and perhaps unpredictable character. Whether or not this characteristic is significant or not requires some appreciation of the manner in which structures respond to dynamic or time-varying loads.

4.2.1. Spring–Mass–Dashpot Systems

Most but certainly not all structures can be approximated by a simple system comprising a spring or elastic restraint, a mass and a dashpot or damper. An understanding of the behaviour of this simple system under dynamic loading is, for the most part, a sufficient base for the understanding of the dynamic behaviour of real structures.

In the system illustrated in Fig. 4.1 the spring is elastic and exerts a force on the mass proportional to the displacement of the mass, i.e.

$$F_{t\,\mathrm{spring}} = -K.x_t \qquad (4.1)$$

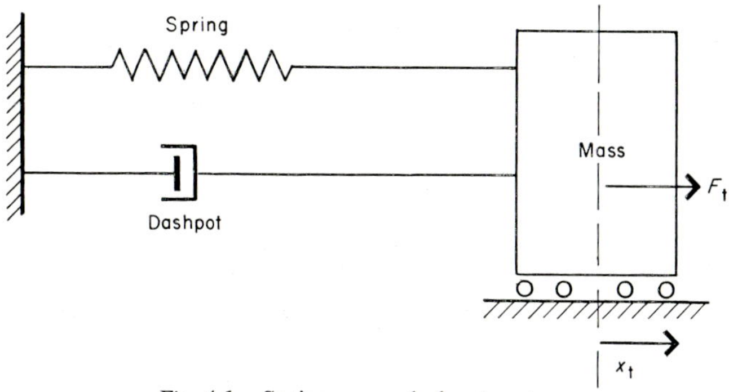

Fig. 4.1. Spring–mass–dashpot system.

where K is the spring stiffness. The sign of the force is negative since it acts in the opposite sense to the displacement, x_t. The spring is a simple representation of the elastic stiffness of a real structure.

The dashpot in Fig. 4.1 is such that it develops a force proportional to the velocity of the mass,

$$v_t = \frac{d}{dt}(x_t) = \dot{x}_t$$

The force exerted by the dashpot on the mass is given by:

$$F_{t\,\text{dashpot}} = -C\dot{x}_t \tag{4.2}$$

where C is the dashpot or damping constant. By its nature the dashpot dissipates energy during displacement and its represents in this simple system the same property in the real structure. In real structures energy is dissipated primarily by material hysteresis and friction but for most purposes it is possible to replace the complex real mechanisms of energy dissipation by an equivalent viscous damper.

The resultant force, $F_{R,\,t}$, acting on the mass will produce an acceleration of the mass equal to $F_{R,\,t}/M$ and hence,

$$\begin{aligned} F_{R,\,t} &= F_t + F_{t\,\text{spring}} + F_{t\,\text{dashpot}} \\ &= M \times \text{acceleration} \\ &= M\frac{d^2x_t}{dt^2} \\ &= M\ddot{x}_t \end{aligned}$$

It then follows that,

$$M\ddot{x}_t + C\dot{x}_t + Kx_t = F_t \tag{4.3}$$

Equation (4.3) is the differential equation of motion of a spring–mass–dashpot system under the action of an external force F_t which varies with time.

4.2.2. *Free Vibrations*

If the external force is zero and the mass is displaced by some amount x_0 and then released at time $t = 0$ the resulting motion is termed a free vibration and it can be shown [4.2] that this motion is given by:

$$x_t = x_0 e^{-2\pi n_0 \zeta t} \cos 2\pi n_1 t \tag{4.4}$$

where n_0 = undamped natural frequency $= 1/2\pi(K/M)^{\frac{1}{2}}$; n_1 = damped natural frequency $= n_0(1-\zeta^2)^{\frac{1}{2}}$; ζ = damping expressed as a fraction of the so-called critical damping $= C/2(MK)^{\frac{1}{2}}$.

Typical values of ζ are given in Table 4.1 and for values of this order there is no significant difference between the damped (n_1) and the undamped (n_0) natural frequencies. An alternative definition of damping is the so-called logarithmic decrement which is defined as the natural logarithm of the ratio of the amplitudes of successive

Table 4.1

Typical damping values

Description of structure and stress level	Damping as a fraction of critical
(i) Low stress levels	
steel frame	0·005–0·010
reinforced prestressed concrete	0·005–0·010
(ii) Working stress	
steel frame	0·010–0·020
reinforced concrete	0·015–0·030
(iii) Near yield	
steel	0·04 –0·06
reinforced concrete	0·05 –0·10

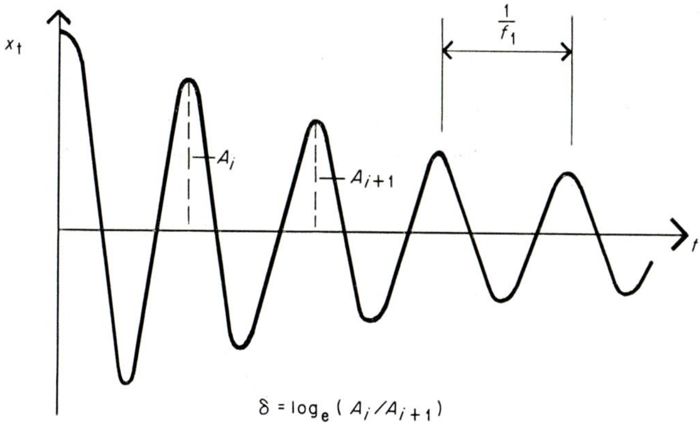

Fig. 4.2. Damped free vibration.

cycles of the free vibration. This is demonstrated in Fig. 4.2 which is a graphical representation of eqn. (4.4). For small values of damping the logarithmic decrement, δ, and the damping as a fraction of critical, ζ, are related as follows:

$$\delta = 2\pi\zeta \tag{4.5}$$

4.2.3. *Sinusoidal Forcing*

If the external force in Fig. 4.1 is sinusoidal in form, i.e.

$$F_t = F_0 \cos 2\pi nt \tag{4.6}$$

then the response can be expressed as:

$$x_t = \frac{F_0}{K} A_{n/n_0,\zeta} \cos (2\pi nt + \phi)$$

where F_0/K = the static deflection or the deflection induced by a static force equal to F_0; $A_{n/n_0,\zeta}$ = a dynamic amplification factor which depends on the frequency ratio, n/n_0, and the damping, ζ; ϕ = a phase angle difference between the force F_t and the response x_t.

The dynamic amplification factor is given by:

$$A_{n/n_0,\zeta} = \left\{ \frac{1}{\left[1-\left(\frac{n}{n_0}\right)^2\right]^2 + 4\zeta^2\left(\frac{n}{n_0}\right)^2} \right\}^{\frac{1}{2}} \tag{4.7}$$

$$= \frac{1}{2\zeta} \text{ when } n = n_0 \tag{4.8}$$

and is plotted in Fig. 4.3. There are a number of features in Fig. 4.3 which deserve particular attention:

(i) if the forcing frequency is less than the natural frequency (say $n/n_0 < 0{\cdot}5$) the dynamic amplification factor is very nearly unity, the response will follow the force in an almost quasi-static fashion with the displacement being very nearly equal to F_t/K;

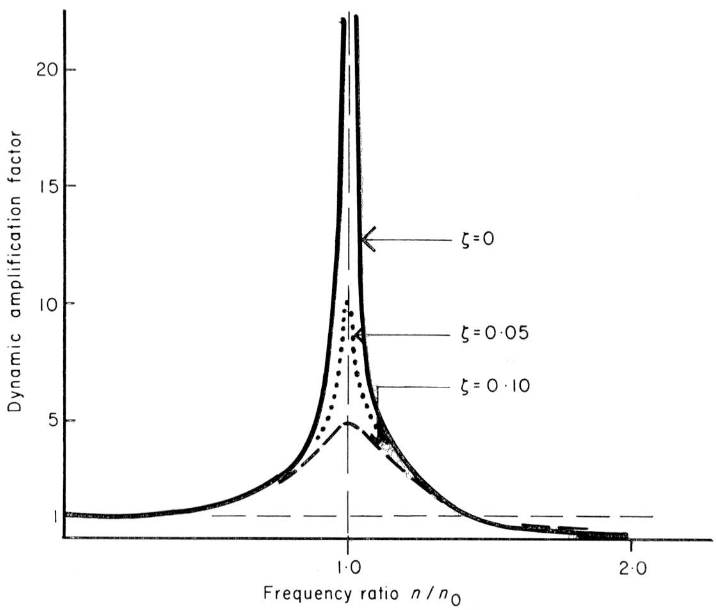

Fig. 4.3. Dynamic amplification factor.

(ii) if the forcing frequency is greater than the natural frequency (say $n/n_0 > 2$) the response is less than the static response of F_t/K, the system is not capable of following the rapid variations in load; and

(iii) in the vicinity of $n = n_0$ the dynamic amplification is substantial and reaches a maximum value of $1/2\zeta$ when $n = n_1 \simeq n_0$.

4.2.4. *Random Forcing*

Generally wind induced forces are not periodic although there are exceptions which are discussed later. Wind forces can usually be described as broad-band random, that is they have a spectrum which extends over a very wide range of frequencies. The response of a structure to wide band forcing can be described qualitatively as follows.

(i) At frequencies lower than n_0 the structural displacement will follow the force in a quasi-static manner;

(ii) the structure will not respond appreciably to frequency components substantially higher than the natural frequency, n_0; and

(iii) the structure will respond strongly to frequency components near the natural frequency.

Figure 4.4 depicts a typical force spectrum. The significance of the so-called 'resonant peak' of the displacement spectrum is dependent upon its area and therefore upon the damping and magnitude of the force spectrum in the vicinity of the natural frequency, n_0. Some typical responses to random forcing are illustrated in Fig. 4.5. Figure 4.5(a) shows a force history with the force expressed as F_t/K. If the natural frequency were infinite or at least well beyond the tail of the force spectrum the response would be quasi-static and its history would be identical to that of F_t/K. For a somewhat lower frequency but a frequency well into the tail of the force spectrum the resonant response will be apparent in the trace, as in Fig. 4.5(b), but will not add significantly to the maximum displacement. If the natural frequency is lower again and into a region of the force spectrum where the energy is appreciable the resonant response will be marked and will increase significantly the peak response (Fig. 4.5(c)). At even lower frequencies the resonant component will be completely dominant and the response trace will appear as a sine wave of

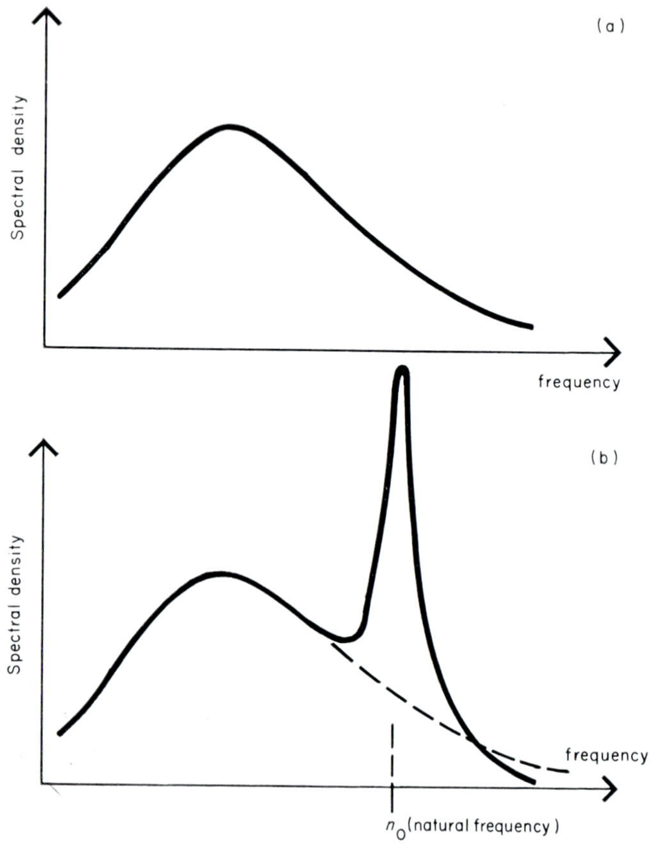

Fig. 4.4. Typical (a) force and (b) response or displacement spectra.

frequency n_0 and an amplitude which varies slowly in a random fashion.

For a quantitative treatment of the response of structures to random forces the reader is referred to the work of Robson [4.3] or Crandall and Mark [4.4].

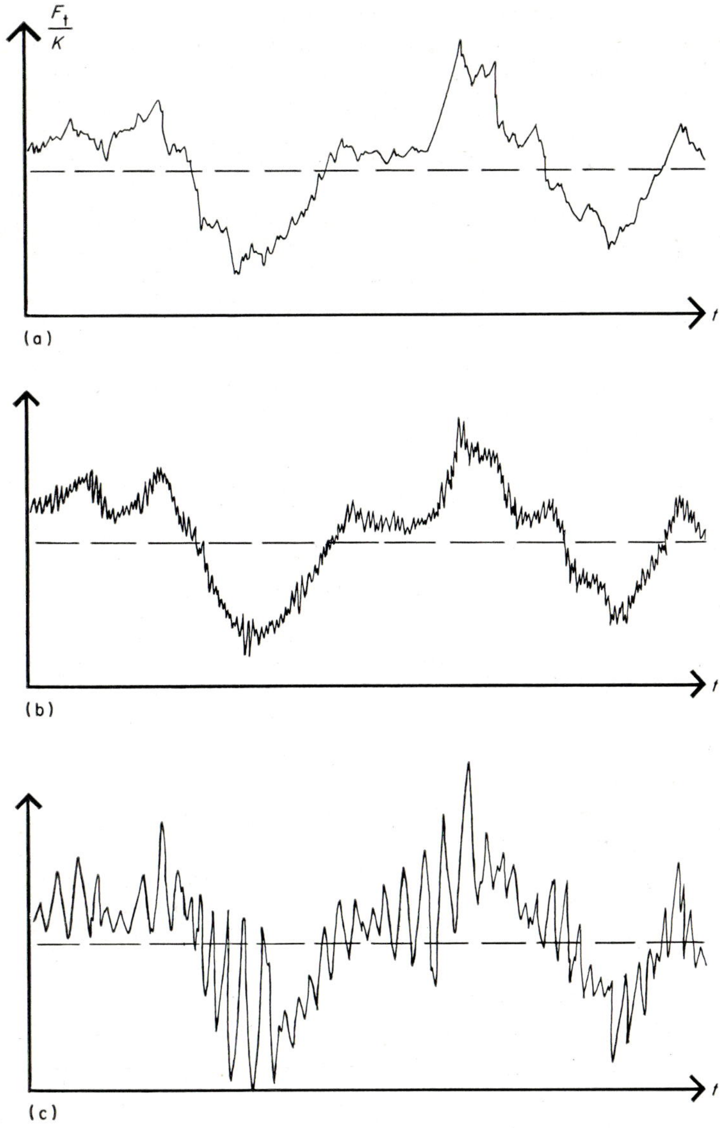

Fig. 4.5. Typical force and response histories. (a) Force time history. (b) Response time history with n_0 in 'tail' of spectrum. (c) Response time history for lower n_0, nearer the energy containing range of force spectrum.

4.3. Design Wind Speed

The initial steps in most codified approaches to the evaluation of wind loads are concerned with the determination of a design speed at the site of the structure. The steps normally involved are:

(i) the selection of a return period;
(ii) the determination of a basic or standard speed corresponding to the selected return period; and
(iii) the adjustment of the standard speed to account for the height, the terrain and the topography of the particular site in question.

4.3.1. Return Periods

At the present time most wind load codes are concerned with the specification of working loads rather than ultimate loads. That is the codes aim to give loads that are likely to occur during the life of the structure and which the structure is expected to resist without signs of distress. This is not the only approach possible and increasing attention is now being given to the specification of ultimate loads or loads which the structure could resist but with a bare safety margin and not without showing signs of failure. Clearly the return periods selected will differ markedly between the two approaches and the remarks here are concerned with the more common working load approach.

Return periods specified in most codes range from 25 years to 50 years with provisions for choosing larger or smaller values for special structures. For temporary structures, return periods as low as a few years might be acceptable while for monumental structures or structures providing essential services, e.g. hospitals and communication centres, a return period of 100 or 200 years might be specified.

For considerations other than strength it may be prudent to choose other return periods although few codes draw attention to this problem. The need to consider other return periods is discussed further in Section 4.9.4 which deals with wind induced accelerations.

4.3.2. Standard or Basic Design Speeds

A standard or basic design speed corresponding to a given return period is that which would be measured under specified conditions of:

(i) terrain;
(ii) topography;
(iii) height; and
(iv) the period over which the speed is averaged.

Speeds specified for standard conditions tend to vary slowly with location and maps depicting variations in the design speed over large

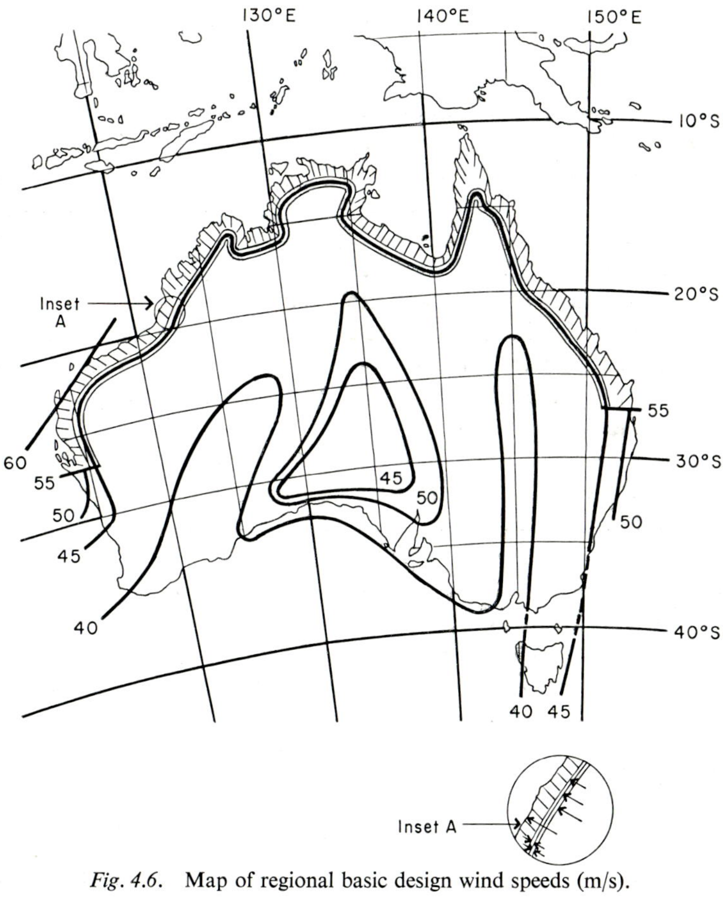

Fig. 4.6. Map of regional basic design wind speeds (m/s).

areas are feasible. Such a map depicting the basic design speeds in Australia [4.5] for a fifty year return period is shown in Fig. 4.6. In this instance the basic speed is defined for:

(i) open country with scattered obstructions (Category 2);
(ii) flat or gently rolling topography;
(iii) a height of 10 m; and
(iv) an averaging period of 2–3 s.

The definition of the basic speed in the U.K. is identical with that given above, while in the U.S.A. the averaging period is a spatial average (the 'fastest mile') which corresponds roughly to a time average of 40 s. The choice of a standard condition is more closely related to the available data than to the advantages that a particular specification might offer. In the U.K. and Australia peak gusts (2–3 s duration) have been recorded over long periods and the data is readily available in a suitable form for analysis. In the U.S.A. the data most readily and most widely available are records of the 'fastest mile of wind'. In Canada the basic speed is the mean hourly value. In terms of application there are arguments for both a 'peak gust' speed, which is particularly suited to the determination of loads on small structures, and the 'mean hourly' speed which is of particular value when the separation of the dynamic and steady loads is desirable.

Irrespective of how the standard speed is defined it is possible, through the use of the mean and gust speed profiles presented in Chapter 3, to adjust the standard to another basis.

4.3.3. Local Design Speeds

The adjustment of the basic speed to a local design speed involves an assessment of the local terrain and topography and the use of wind speed profiles presented in Chapter 3. This procedure is best illustrated by example.

4.3.3.1. Example 1

Given that the basic design speed for a 2–3 s gust at 10 m over flat open country is 50 m/s determine the mean hourly speed and the peak gust speed at a height of 50 m in a suburban area (Category 3).

From eqn. (3.9): $\hat{V}(10)/\hat{V}(z_G) = (10/z_G)^{\beta}$ and from Table 3.1: $z_G = 300$ m and $\beta = 0{\cdot}09$; hence $\hat{V}(z_G) = 50 \times (300/10)^{0{\cdot}09} = 68$ m/s; and from the eqn. (3.9a): $\overline{V}(z_G) = \hat{V}(z_G)/1{\cdot}35 = 50$ m/s.

Having defined the gradient speeds by extrapolating upwards from 10 m using the profile for Category 2 (flat open country), the speeds in Category 3 at 50 m are determined as follows:

$$\frac{\hat{V}(50)}{\hat{V}(z_G)} = \left(\frac{50}{400}\right)^{0{\cdot}14}$$

and

$$\frac{\overline{V}(50)}{\overline{V}(z_G)} = \left(\frac{50}{400}\right)^{0{\cdot}25}$$

hence $\hat{V}(50) = 51$ m/s and $\overline{V}(50) = 28$ m/s.

4.4. Design Wind Loads

Having determined a design wind speed there are a number of possible approaches to the determination of the design wind loads. These approaches vary in complexity and depth from a detailed dynamic analysis coupled with wind tunnel testing to the simple quasi-static approach which is the basis of most wind codes.

4.4.1. Quasi-static Methods

In quasi-static (or quasi-steady) approaches to the evaluation of wind loads it is assumed that the maximum loads imposed during a storm are induced by the maximum gust and that these loads can be computed directly from pressure or load coefficients as defined in Chapter 2 and used in conjunction with a 'velocity pressure' equal to $\frac{1}{2}\rho\hat{V}^2$ where $\hat{V}$ is the peak gust speed. The major deficiencies of the approach are:

(i) no account is taken of the dynamic response of the structure;
(ii) in its basic form, no account is taken of the size of the building relative to gust size; and
(iii) the pressure coefficients and load coefficients employed are commonly those determined from measured mean loads and mean speeds.

Notwithstanding these deficiencies the method yields acceptable estimates of load for a wide range of structures. An appreciation of the type of structure for which the method is likely to be inadequate can be obtained by considering the deficiencies outlined above in further detail.

It has been demonstrated that frequency components up to about one half of the natural frequency produce a quasi-static response, i.e. the deflection is very nearly equal to the applied force divided by the stiffness. Above one-half of the natural frequency there is some amplification of the load effect but large amplifications occur only in the vicinity of a natural frequency. It follows therefore that dynamic effects will only be significant if the natural frequency lies well into the tail of the load spectrum and the response is as illustrated in Fig. 4.5(a) or Fig. 4.5(b). The load spectrum is commonly similar in form to the turbulence spectrum shown in Fig. 3.12 but falls away more quickly at the high frequency end or tail. The fall-off is associated with the inability of the smaller, high frequency gusts to completely envelope the structure and hence develop the full load potential. The fall-off is such that if the wavelength, λ, is equal to or less than a characteristic dimension of the structure the load spectrum at that wavelength is, for most practical purposes, negligible. Since the wavelength exciting the natural frequency is $\overline{V}/n_0$ it can be deduced that dynamic amplification will be insignificant when $\overline{V}/n_0L < 1$, approximately. L is a characteristic dimension of the structure and may be taken as the square root of the frontal area of the structure. Most structures do in fact fall into this category ($\overline{V}/n_0L < 1$); exceptions are tall buildings, particularly slender tall buildings and unusually flexible structures.

The size effect which is ignored in the simpler quasi-static approaches has been introduced in the consideration of the dynamic response and it is in this area that the effect is most pronounced. In terms of the dynamic or resonant response the significant parameter is $\overline{V}/n_0L$ whereas for the overall or full spectrum gust loads the significant parameter is the ratio λ_E/L where λ_E is a wavelength characteristic of the energy containing region of the spectrum. As shown in Fig. 3.12 λ_E is about 500–1000 m and hence for most structures the size parameter $\lambda_E/L \gg 1$ and reductions in the gust loads due to this effect are small. Reductions in the total load (which includes both the mean and gust components) are, on a percentage basis, smaller again. Size reduction factors are sometimes included as a refinement to the simple quasi-static approach. The Australian code allows reductions in load of up to a maximum of 20% for areas of 15 000 m^2. The U.K. code modifies the gust duration from 2–3 s for small structures to 15 s for structures whose greatest dimension is 50 m or more. This change is equivalent to a load reduction of a little less than 20%.

There is a danger in applying size reduction factors to the quasi-steady approach since the reduction is greatest for large structures and large structures tend to possess low natural frequencies and hence are likely to be subject to a significant resonant response. In most cases the load reductions applied are reasonable but care should be exercised if the calculated natural frequency is lower than about 0·5 Hz.

The third deficiency of the quasi-steady approach is more a fault of present practice rather than the approach itself. Most of the pressure coefficients employed in codes of practice have been derived from measurements of mean load and mean pressure, i.e. the coefficient, $\bar{C}_p$, has been determined from the relationship:

$$\bar{C}_p = \frac{\bar{p}}{\frac{1}{2}\rho\bar{V}^2} \tag{4.9}$$

where $\bar{p}$ = measured mean pressure and $\bar{V}$ = measured mean speed.

The value of $\bar{C}_p$ so determined is then used to determine a *peak* pressure, $\hat{p}$, using

$$\hat{p} = \bar{C}_p\tfrac{1}{2}\rho\hat{V}^2 \tag{4.10}$$

Clearly, the pressure coefficient employed in eqn. (4.10) should be determined from measurements of $\hat{p}$ and $\hat{V}$, i.e. a coefficient, $\hat{C}_p$, defined as:

$$\hat{C}_p = \frac{\hat{p}}{\frac{1}{2}\rho\hat{V}^2} \tag{4.11}$$

should be employed.

Unfortunately $\hat{C}_p$ and $\bar{C}_p$ as defined in eqns. (4.9) and (4.11) are not identical. Values of $\hat{C}_p$ and $\bar{C}_p$ determined in a full scale investigation by Eaton and Mayne [4.6] are presented in Fig. 4.7. The

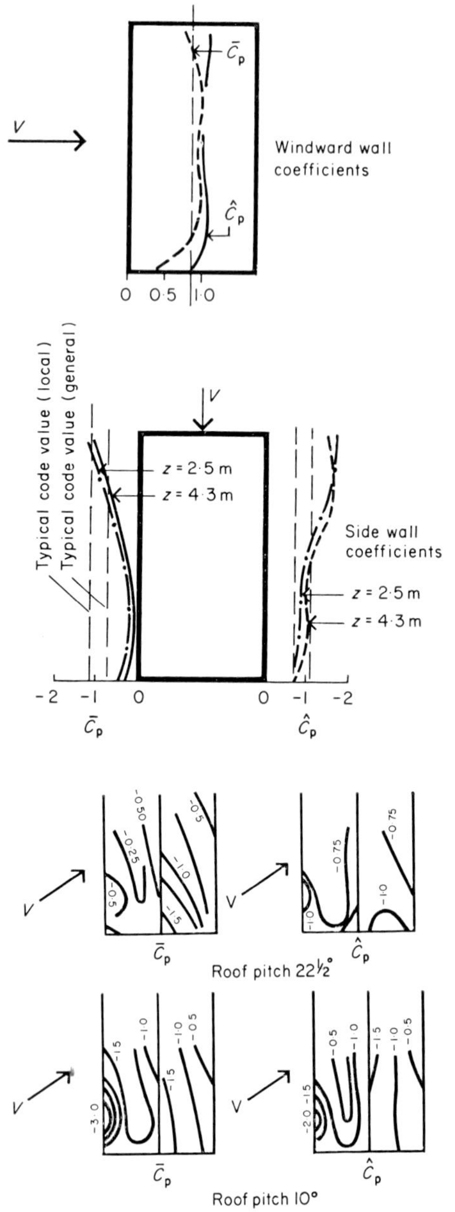

Fig. 4.7. Full scale measurements of $\overline{C}_{\rm p}$ and $\hat{C}_{\rm p}$ (Eaton and Mayne [4.6]). Coefficients based on 5 m speeds at eaves height.

discrepancies are in many cases alarming but fortunately there is fair agreement for the more important results such as the total load on the front wall and, to a lesser degree, the total roof load.

Further results, in this case wind tunnel observations of the loads on roof segments of a large hangar, are presented in Fig. 4.9. Figure 4.9 shows coefficients $\hat{C}_p$ and $\bar{C}_p$ for a segment of the roof of the model shown in Fig. 4.8; the coefficients are for the total load, due

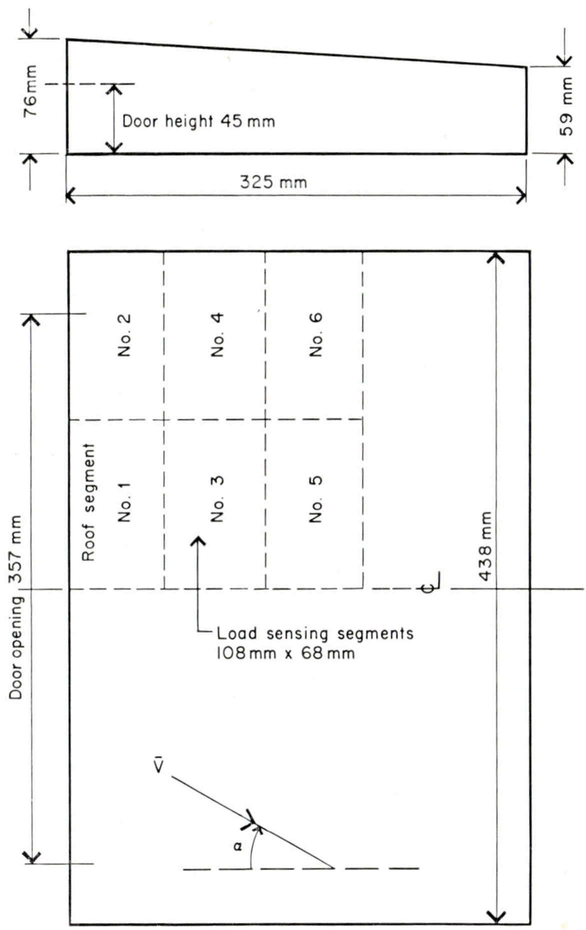

Fig. 4.8. Model dimensions (scale 1:300).

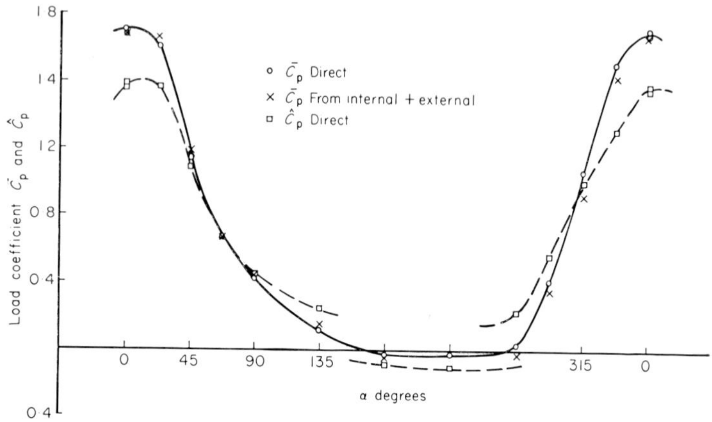

Fig. 4.9. Load coefficients for segment No. 2 doors fully open.

to both internal and external pressures, acting on the segment. For the critical case the values of $\bar{C}_p$ and $\hat{C}_p$ differ by about 20% and discrepancies of this order are not uncommon.

4.4.2. Gust Factor Methods

In recent years attempts have been made to place the evaluation of wind loads on a more rational basis. The so-called 'gust factor' approach introduced by Davenport [4.7] and Vickery [4.8] and developed by Vickery [4.9] and Simiu [4.10 and 4.11] is one such method. The essential feature of the method is the separation of the mean and fluctuating components of wind load. The mean loads are determined from the mean wind speed using pressure or load coefficients as defined in eqn. (4.9). The fluctuating loads are determined separately by a method which makes a rational allowance for the intensity of turbulence at the site, size reduction effects and dynamic amplification. In most instances the estimates of peak load obtained from the gust factor approach are very similar to those obtained using the more conventional quasi-steady method. The prime advantages of the gust factor method are:

(i) the ability to detect situations in which dynamic effects will be significant; and
(ii) an improved description of the nature of the loading.

The second point is of considerable significance since the quasi-steady approach tends to suggest a single static load rather than a load history typical of those shown in Fig. 4.5. The gust factor approach yields quantitative information concerning:

(i) the magnitude of the mean load;
(ii) the magnitude of the comparatively slow variations in load associated with the larger gusts; and
(iii) the magnitude of the 'resonant' component.

Information concerning the nature of the load history is invaluable in assessing the possibility of fatigue damage as opposed to damage caused by a single overload.

4.4.3. Wind Tunnel Methods

Neither the quasi-steady nor the gust factor approach to the assessment of wind load can be relied upon to produce load estimates with an accuracy of better than about 20% and for unusual structures

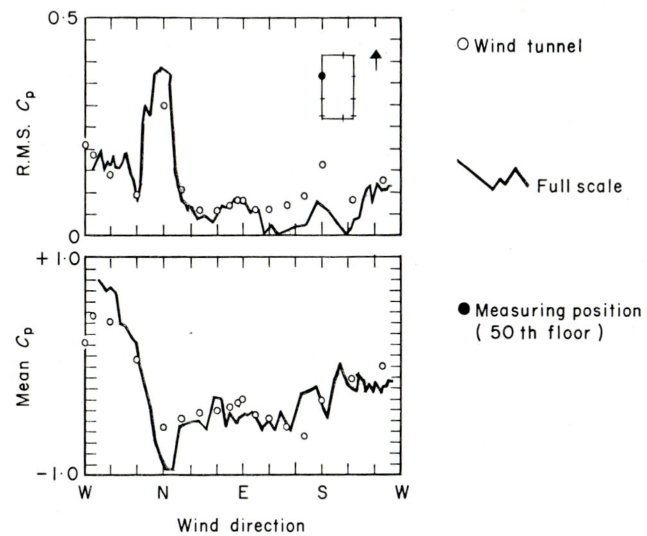

Fig. 4.10. Comparison of wind tunnel and full scale measurements of surface pressure coefficients on the wall of a tall building (after Dalgleish [4.12]).

the accuracy would be poorer due to the lack of reliable pressure coefficients. In the case of unusual structures and also for structures involving large investments it is advisable to determine wind loads by means of model tests. Correctly scaled wind tunnel tests presently offer the most reliable means for the determination of wind loads and wind effects. It is essential, however, that such tests be correctly scaled, as detailed in Chapter 2. Tests conducted in smooth uniform flow (and prior to 1964 or thereabouts, most wind tunnel tests were conducted in this type of flow) can yield very misleading results. If, however, the tests are conducted in a boundary layer tunnel and the model is scaled to match the model atmosphere of the tunnel the results will be a fair representation of reality. Recent comparisons, e.g. [4.12] of full scale measurements with wind tunnel tests results have proved most encouraging. A comparison of some full scale and wind tunnel observations are shown in Fig. 4.10.

4.5. Wind Load Calculations

The calculations of wind loads by the quasi-steady approach is treated in considerable detail in Reference 4.1 and the treatment here is brief and is restricted to comparatively simple and common building shapes.

4.5.1. External Pressure Distribution

A simplified pressure distribution on a typical low rise dwelling or industrial building is shown in Fig. 4.11. The pressure coefficients shown in Fig. 4.11 are essentially those recommended in the Australian Wind Code but similar values are prescribed in most codes. Even for such a simple structure there are four variables (h/d, b/D, θ, wind profile) and a complete specification of the pressure distribution is not feasible particularly if variations in wind direction are to be considered. It is the aim of codes to specify loads simply but at the same time reasonably and in these circumstances departures from reality are unavoidable.

Distributions such as shown in Fig. 4.11 can only be fair approximations and this should be borne in mind when they are employed in the design process. In this example care must be taken when

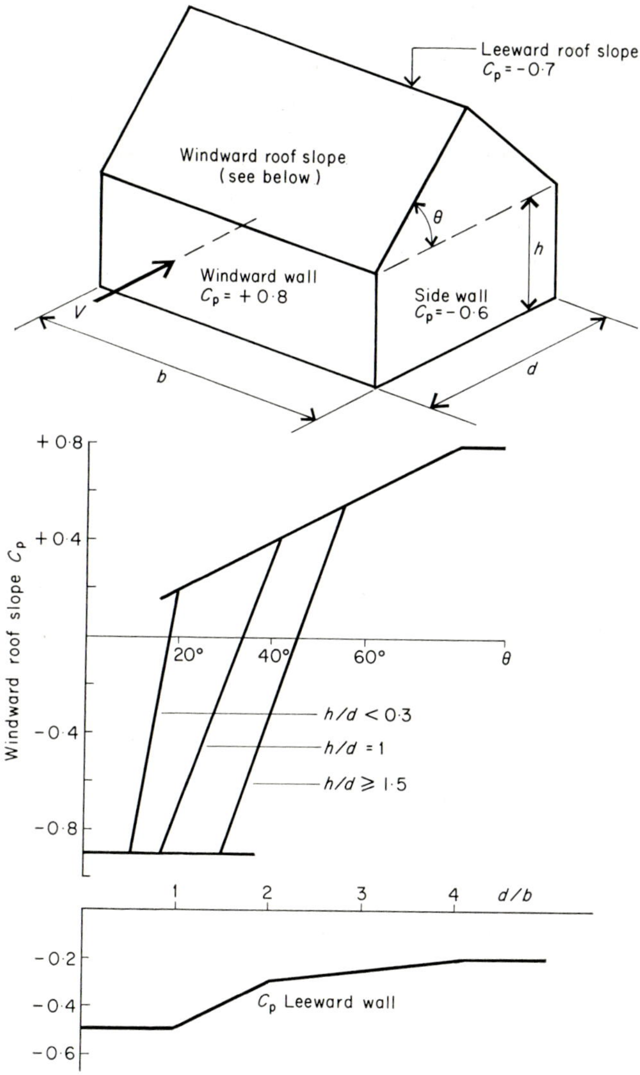

Fig. 4.11. Typical code pressure coefficients for a pitched roof shed.

determining roof loads which it can be seen, are extremely sensitive to roof slope, and it would be advisable to consider a reasonable range of loads rather than a single value. For example, at low values of h/d and roof slopes between 10° and 20° there is a rapid change from a C_p of $-0{\cdot}9$ to $+0{\cdot}2$; even near the ground the wind direction is not constantly in a horizontal plane and the effective roof slope will vary considerably and any coefficient within the stated range should be considered and the worst case adopted.

For low rise structures the pressure coefficients are employed together with the speed at eaves level and the pressure assumed to be constant over the face in question. For taller structures the variation in wind speed with height is significant and must be taken into account. The variation is most marked on the front face although even within the wake region some variation with height is apparent. The procedure commonly adopted in Codes is to compute frontal pressures using a constant pressure coefficient with a speed which varies with height in accordance with the gust profiles specified in Chapter 3 but to accept constant pressures within the wake (roof, side walls and rear wall). The wake pressures are computed from the relevant pressure coefficient and the wind speed at eaves level or roof level.

4.5.1.1. Example 2

Compute the total drag (horizontal) force and total lift (vertical) force on the building shown in Fig. 4.12.

Data

Basic design speed (2–3 s gust at 10 m) in Category 2	45 m/s
Terrain type	Cat. 2
Wind direction	As shown in Fig. 4.12

Computation

Local design speed (eaves height 5 m)

$$V\,(5\text{ m}) = 45 \times \left(\frac{5}{10}\right)^{0.09}$$

$$= 42{\cdot}3\text{ m/s}$$

Design pressure

$$= \tfrac{1}{2}\rho V^2$$

$$= 0{\cdot}6 \times 42{\cdot}3^2\text{ Pa}$$

$$= 1{\cdot}07\text{ kPa}$$

Horizontal load on Face 1

$$C_p = +0{\cdot}8$$
$$p = 0{\cdot}8 \times 1{\cdot}07 \text{ kPa}$$
$$F_1 = pA$$
$$= 0{\cdot}8 \times 1{\cdot}07 \times 20 \times 5 \text{ kN}$$

Horizontal load on Face 4

$$C_p = -0{\cdot}3$$
$$F_4 = 32 \text{ kN}$$

Loads on Face 2

$$C_p = -0{\cdot}9$$
$$F_4 \text{ (horizontal)} = 0{\cdot}9 \times 1{\cdot}07 \times 20 \times 1{\cdot}8$$
$$= 35 \text{ kN}$$
$$F_4 \text{ (vertical)} = 0{\cdot}9 \times 1{\cdot}07 \times 20 \times 5$$
$$= 96 \text{ kN}$$

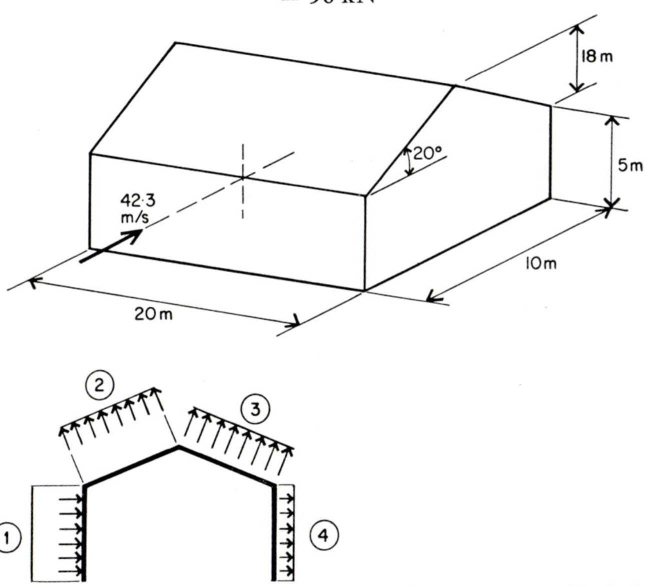

Fig. 4.12. Example 2. Basic design speed 45 m/s (2–3 s gust at 10 m). Local design speed (Cat. 2 terrain) = 42·3 m/s. Total loads: Surface 1, 86 kN; Surface 2, 102 kN; Surface 3, 80 kN; Surface 4, 32 kN.

Loads on Face 3 $(C_p = -0{\cdot}7)$

$F_3 = 27$ kN (horizontal)

$F_3 = 75$ kN (vertical)

With due regard to direction

Total drag force $= 86+32-35+27$

$= 110$ kN

Total lift force $= 96+75$

$= 171$ kN

4.5.1.2. Example 3

Compute the base shear force and the overturning moment about the base for the building shown in Fig. 4.13.

Data:

Basic design speed (2–3 s gust at 10 m) in Category 2	45 m/s
Terrain type	Cat. 4
Wind direction	as shown in Fig. 4.13

Computations

Basic design speed (10 m)	45 m/s
2–3 s gust at gradient height in Cat. 2	$45\times\left(\frac{300}{10}\right)^{0.09}$
	$= 61{\cdot}1$ m/s
Local design speed 2–3 s gust at 50 m in Cat. 4	$= 61{\cdot}1\times\left(\frac{50}{500}\right)^{0.20}$
	$= 38{\cdot}6$ m/s
Design pressure	$= 0{\cdot}6\times 38{\cdot}6^2$ Pa
	$= 0{\cdot}89$ kPa

Load on rear face

$C_p = -0{\cdot}5$

$p = 0{\cdot}5\times 0{\cdot}89$ kPa

Total force $= 0{\cdot}5\times 0{\cdot}89\times 50\times 60$ kN

$= 1335$ kN

Base moment $= 1335\times 60/2$

$= 40\,050$ kN m

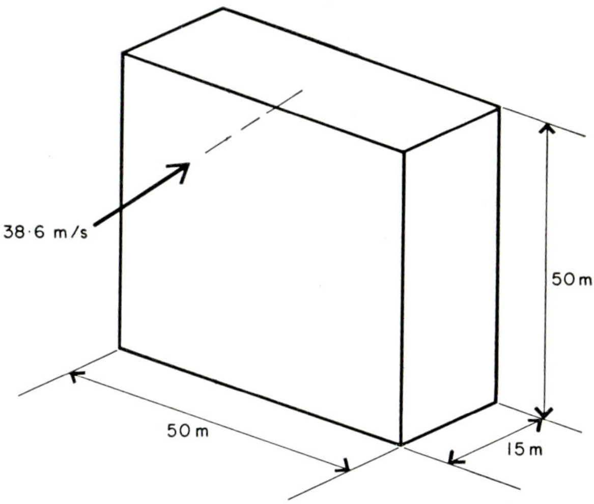

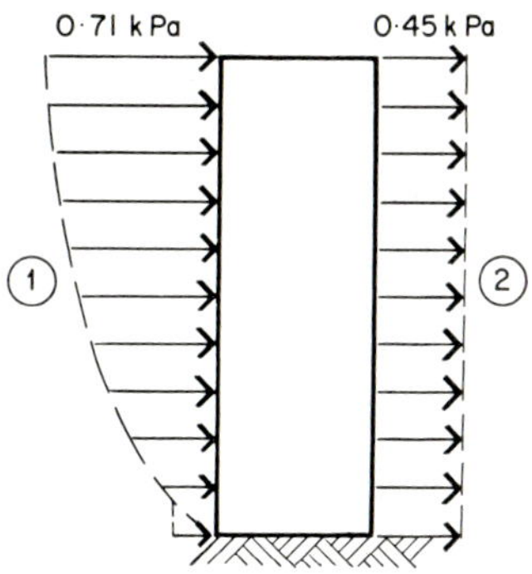

Fig.. 4.13. Example 3. Total force: Face 1, 1533 kN; Face 2, 1335 kN (=2868 kN).

Loads on front face

For buildings over about 20 m in height the frontal pressures are normally calculated with an allowance for the increase in speed with height. In the present example

$$\frac{v(z)}{v(60)} = \left(\frac{z}{60}\right)^{0.2}$$

and

$$\frac{q(z)}{q(60)} = \left(\frac{z}{60}\right)^{0.4}$$

or,

$$q(z) = 0{\cdot}89 \left(\frac{z}{60}\right)^{0.4}$$

$$\begin{aligned} p(z) &= C_p q(z) \\ &= 0{\cdot}80 \times 0{\cdot}89 \times \left(\frac{z}{60}\right)^{0.4} \\ &= 0{\cdot}71 \left(\frac{z}{60}\right)^{0.4} \text{ kPa} \end{aligned}$$

The total load can be calculated by integration or by a summation of the loads on say 4 or 5 subdivisions of the height. In the present case the loads and moments are calculated by dividing the building into 5 segments each of 12 m. The calculations are presented in Table 4.2.

Table 4.2

Front face loads and moments

Segment No.	Height range (m)	Average ht (m)	$\bar{p}$ (kPa)	Load (kN)	Moment (kN m)
1	0–12	6	0·28	170	1 018
2	12–24	18	0·44	263	4 737
3	24–36	30	0·54	323	9 685
4	36–48	42	0·62	369	15 513
5	48–60	54	0·68	408	22 055
			Σ	1 533	53 010

In detail, the calculations for segment 3 are:

Average height $= 30$ m

$p = 0{\cdot}71 \times \left(\frac{30}{60}\right)^{0.4}$ kPa

$= 0{\cdot}54$ kPa

Load $= 0{\cdot}54 \times 12 \times 50$

$= 323$ kN

Moment $= 323 \times 30$

$= 9685$ kN m

finally: Total shear force $= 1335 + 1533$

$= 2868$ kN

Base moment $= 40\,050 + 53\,010$

$= 93\,060$ kN m

4.5.2. Internal Pressures

All buildings are vented in one way or another to the atmosphere and during strong wind storms this venting will, in general, produce internal pressures differing from the free stream static pressure. The magnitude of the internal pressure depends primarily on the distribution of vents or openings in relation to the external pressure distribution. With dominant venting on a front face the internal pressure will be positive while for dominant venting within the wake region the internal pressure will be negative.

In a steady state the internal pressures can be computed from a knowledge of the external pressure distribution and the size, shape and distribution of the vents or openings. In the steady state (mean or time average) the inflow and outflow must balance. The flow through a wall subject to some uniform external pressure p_e, an internal pressure p_i and with a total area open, A, is given by:

$$Q = C_d A \left(\frac{2}{\rho}(p_e - p_i)\right)^{\frac{1}{2}} \tag{4.12}$$

with the direction of flow being from the higher pressure. If the pressures are expressed in terms of coefficients, i.e.

$$p_e = C_{p_e} . \frac{\rho v^2}{2}$$

then

$$Q = C_d A v (C_{p_e} - C_{p_i})^{\frac{1}{2}} \tag{4.13}$$

where C_d is a discharge coefficient which depends primarily on the shape of the opening or leakage path. The value of C_{p_i} can be computed from a knowledge of the values of C_d, A and C_{p_e} for each external surface by use of the equation,

$$\Sigma Q = 0$$

or

$$\Sigma \frac{(C_{p_e} - C_{p_i})}{|C_{p_e} - C_{p_i}|} C_d A \left(|C_{p_e} - C_{p_i}|\right)^{\frac{1}{2}} = 0 \tag{4.14}$$

4.5.2.1. Example 4

Compute the internal pressure coefficient, C_{p_i}, for the building shown in Fig. 4.14; the external pressure coefficients are as shown in Fig. 4.14 and the values of $C_d A$ for each wall are equal.

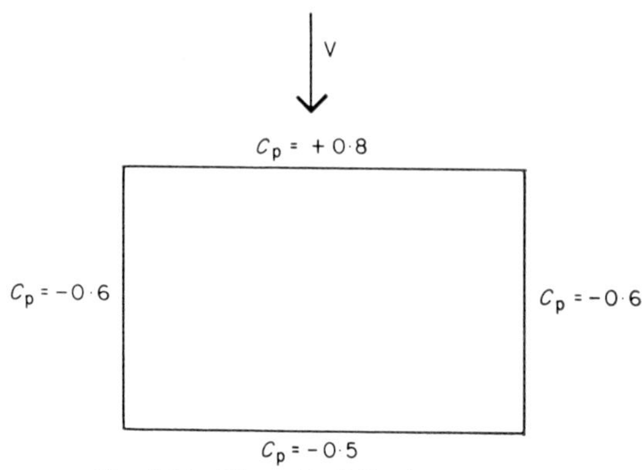

Fig. 4.14. Plan of building in Example 4.

Solution

Equation (4.14) is best solved by trial and error by computing Q for a range of values of C_{p_i} and plotting Q *v.* C_{p_i} as shown in Fig. 4.15. Assuming $C_{p_i} = -0{\cdot}4$, we have:

$$\begin{aligned}\frac{Q}{C_d A} &= (0{\cdot}8-(-0{\cdot}4))^{\frac{1}{2}} - (-0{\cdot}4-(-0{\cdot}6))^{\frac{1}{2}} \\ &\quad - (-0{\cdot}4-(-0{\cdot}6))^{\frac{1}{2}} - (-0{\cdot}5-(-0{\cdot}6))^{\frac{1}{2}} \\ &= (1{\cdot}2)^{\frac{1}{2}} - (0{\cdot}2)^{\frac{1}{2}} - (0{\cdot}2)^{\frac{1}{2}} - (0{\cdot}1)^{\frac{1}{2}} \\ &= -0{\cdot}12\end{aligned}$$

Similarly for $C_{p_i} = -0{\cdot}5$

$$\frac{Q}{C_d A} = (1{\cdot}3)^{\frac{1}{2}} - (0{\cdot}1)^{\frac{1}{2}} - (0{\cdot}1)^{\frac{1}{2}} - (0)^{\frac{1}{2}}$$

$$= +0{\cdot}51$$

and for $C_{p_i} = -0{\cdot}3$

$$\frac{Q}{C_d A} = (1{\cdot}1)^{\frac{1}{2}} - (0{\cdot}3)^{\frac{1}{2}} - (0{\cdot}3)^{\frac{1}{2}} - (0{\cdot}2)^{\frac{1}{2}}$$

$$= -0{\cdot}49$$

The three trial solutions are plotted in Fig. 4.15 and from this plot it is seen that for $\Sigma Q = 0$ the internal pressure coefficient is $-0{\cdot}42$.

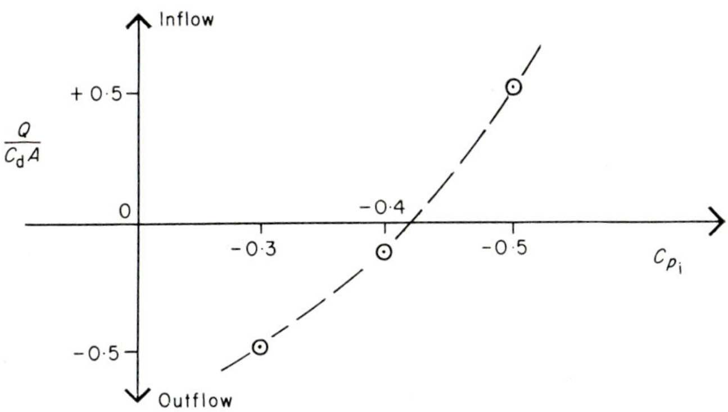

Fig. 4.15. Plot of C_{p_i} *v.* $Q/C_d A$.

An example of wind tunnel observations of internal pressure is shown in Fig. 4.16. The model employed was that shown in Fig. 4.8 with the large hangar door fully open. Figure 4.16 shows the mean pressure coefficient together with both the maximum and minimum values and it is clear, for all wind directions, that the internal pressure fluctuates widely. Except for a small range of wind directions, the maximum and minimum values are of opposite sign. The conventional quasi-steady approach used in most codes of practice will give an indication only of the peak which is of the same sign as the mean value.

Figure 4.17 illustrates the influence of the size of the openings on

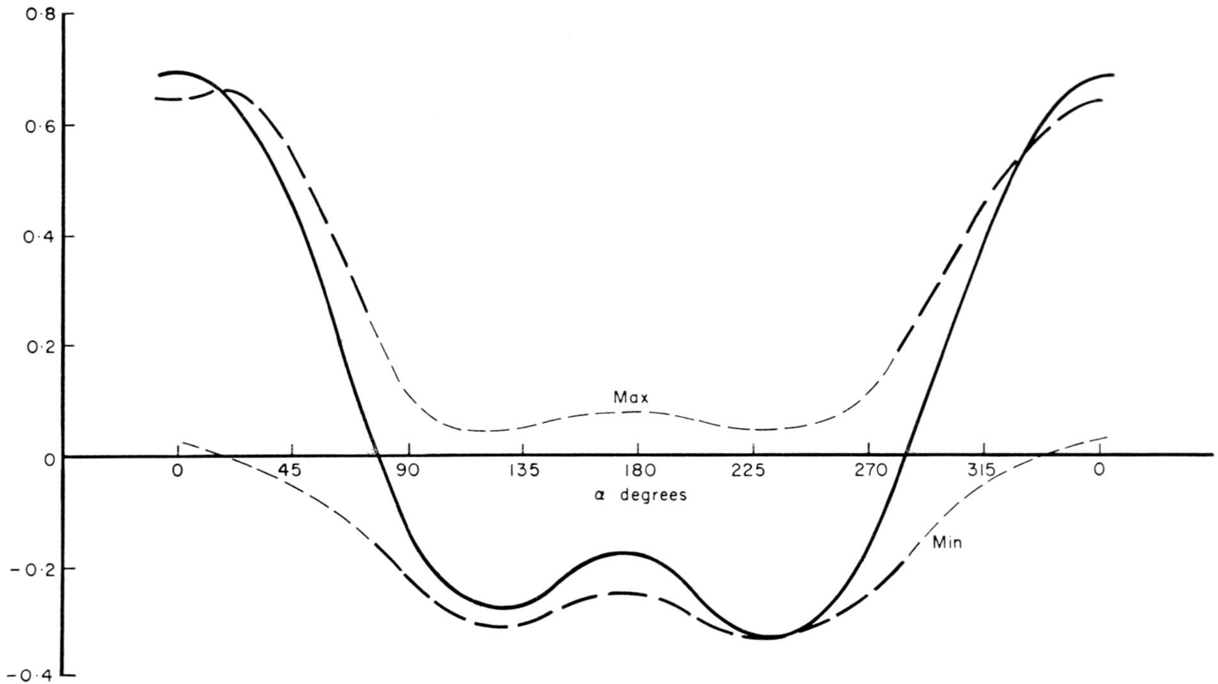

Fig. 4.16. Mean, maximum and minimum internal pressure coefficients. Mean coefficient $=\bar{p}/\frac{1}{2}\rho\bar{u}^2$; peak coefficient $=\hat{p}/\frac{1}{2}\rho\hat{u}^2$ (doors fully open).

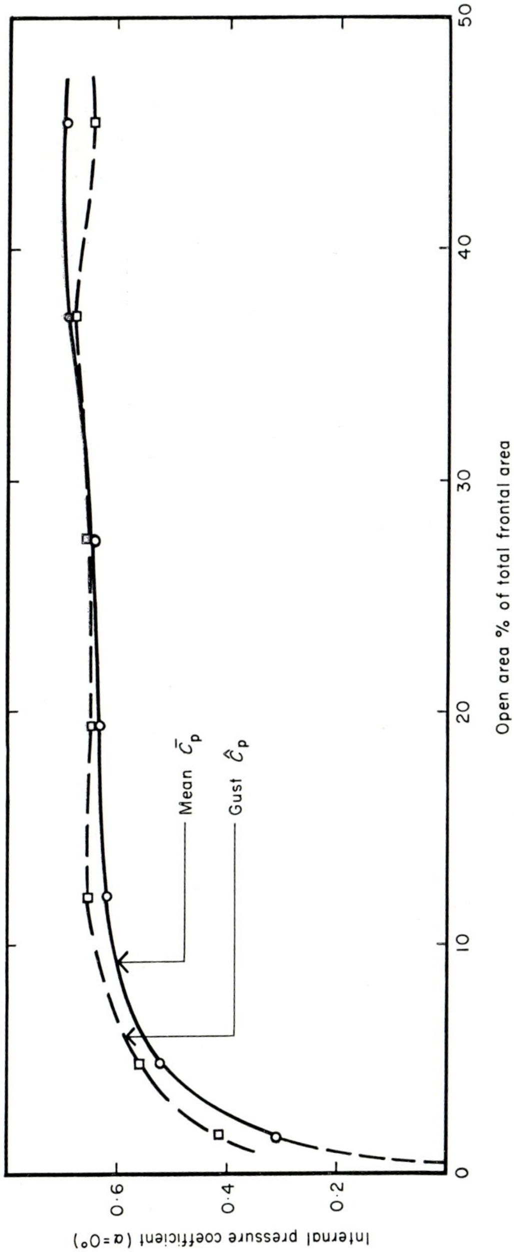

Fig. 4.17. Variation of internal pressure with door opening area. Roof venting ≃0·5% of frontal area.

the internal pressure. For this particular test the roof openings amounted to about $\frac{1}{2}\%$ of the frontal area and the open frontal area was reduced from a maximum of about 45% down to a little less than 2% of the total frontal area. It might be noted that there is no significant change in the internal pressure until the frontal opening is reduced to about 5%. The values of $\hat{C}_p$ and $\bar{C}_p$ are in good agreement but, as is the case in Fig. 4.16, the values diverge as C_{p_i} tends to zero. This is a significant point and it should be noted that where quasi-steady based codes suggest values of C_{p_i} near zero that, due to gusts and particularly the non-uniformity of gusts, the actual pressure coefficient could well fluctuate within a range of typically $\pm 0{\cdot}3$.

4.5.3. Loads on Walls and Roofs

The total or overall load on a complete, closed building is determined entirely by the external pressure distribution. The loads on the external skin, are however determined by the difference between the internal and external pressures. Loads on internal walls and also ceiling loads are caused by variations of internal pressure within the building. In the case of industrial buildings without ceilings or floor to roof partitions the computation of the loads on the external walls and the roof is straightforward, although careful consideration should

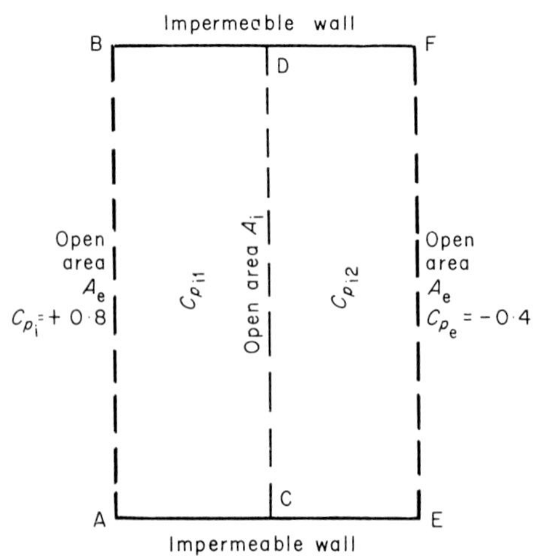

Fig. 4.18. Plan view of building with an internal partition.

be given to possible openings during extreme winds. Doors, particularly large roller doors, are susceptible to failure and, unless specially designed, should be regarded as potential openings. Similarly, windows should be regarded as potential openings unless adequately designed and protected from missile damage.

Wall or roof loads are more difficult to determine when ceilings and full height partition walls are fitted although it is generally possible to make reasonable estimates if steady state conditions are assumed to apply. If the openings in internal partitions and ceilings are large compared with the venting through the exterior walls then their influence can be neglected but, with this position reversed, the internal walls and ceilings will be subject to substantial wind loading.

Figure 4.18 shows the plan of a building with two porous walls and a single full width internal partition. In the steady state situation the flows through AB, CD and EF must be equal and hence:

$$\begin{aligned} Q &= C_d A_e (0{\cdot}8 - C_{p_{i1}})^{\frac{1}{2}} \\ &= C_d A_i (C_{p_{i1}} - C_{p_{i2}})^{\frac{1}{2}} \\ &= C_d A_e (C_{p_{i2}} - (-0{\cdot}4))^{\frac{1}{2}} \end{aligned}$$

The values of $C_{p_{i1}}$ and $C_{p_{i2}}$ may be determined if A_i/A_e is known and the values are shown in Fig. 4.19. At low values of A_i/A_e the partition carries the full pressure difference between the windward and leeward external pressures but for $A_i/A_e > 2$ the partition has little influence on the internal pressure and itself carries only minor wind loads.

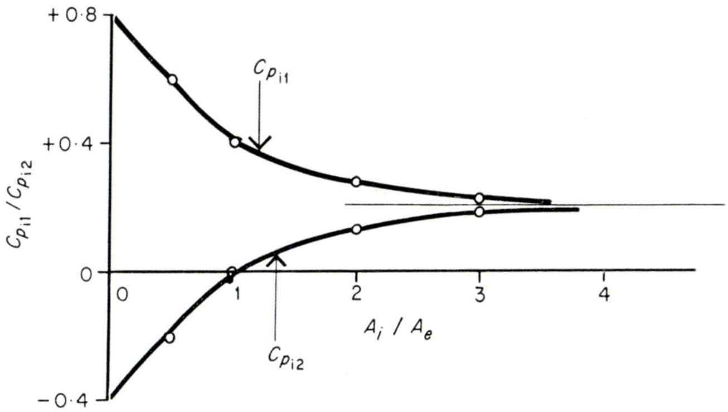

Fig. 4.19. Internal pressure coefficients for the building in Fig. 4.18.

4.5.4. *Local Loads*

The typical 'code' distribution of external pressure shown in Fig. 4.11 is simplified and purports to describe average surface loads only. In some areas of a given surface and perhaps for wind directions other

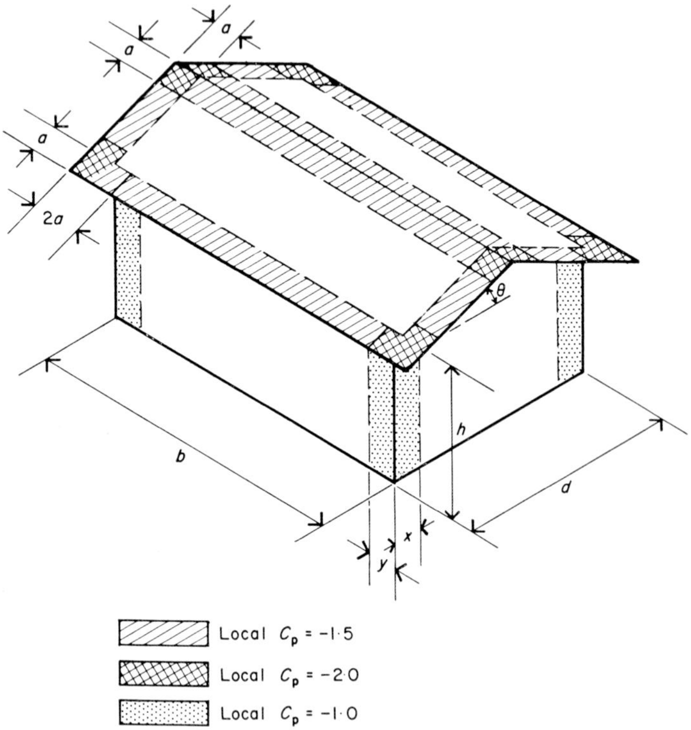

Fig. 4.20. Local pressure coefficients. a = lesser of 0·5 h and 0·1 b; x = 0·10 b; y = 0·10 d.

than that which produces the maximum total load there are commonly pressure coefficients which are significantly more severe than the average shown in Fig. 4.11. These local pressures are generally not important in regard to the overall strength and stability of the building but must be considered in the design of the cladding and the minor structural members in the region.

The difference between the local and the surface average pressures is only significant where the coefficient is negative. Large local 'suctions' commonly occur near surface discontinuities such as the eaves, the ridge or a corner. The more important regions together with suggested pressure coefficients are shown in Fig. 4.20 but other discontinuities, such as chimneys, may produce high local suctions in their vicinity.

4.6. Wake Induced Dynamic Loads

From the discussion of dynamic loads to this point it might have been inferred that these loads were due only to the dynamic nature (gusting) of the wind. However, even in smooth non-turbulent flow the wakes of bluff bodies, such as buildings, are unsteady and fluctuating forces are induced. These forces act primarily in a direction normal to the wind and induce across wind motions. Wake induced fluctuations in drag and also torsion do occur but these are generally small compared with the fluctuations induced directly by gusting. The nature of the wake has been discussed in Chapter 2 but the more important characteristics will be reviewed here.

4.6.1. Spectrum of Across-Wind Forces

Spectra characteristic of across-wind forces are shown in Fig. 4.21. These spectra may be described in terms of their relative breadth (narrow or broad band) and their central frequency. The central frequency (the frequency at which the spectrum attains a maximum) when expressed in the form $nb/\bar{v}$ is is known as the Strouhal number. The Strouhal number S, varies from about 0·1 for rectangular cross-sections to about 0·25 for circular cross-sections at very high values of Reynolds number. The breadth of spectrum increases as:

(i) the turbulence level is increased;
(ii) the after-body length is increased; and
(iii) the aspect ratio (length: breadth) is decreased.

The dynamic response to wake induced forces can be described in terms of the spectrum. At low wind speeds $n_0 b/\bar{v}$ is high and located well to the right of the spectral peak; the available energy is below the natural frequency and the response is quasi-static with no appreciable dynamic amplification. As the wind speed is increased the total energy increases and since the parameter $n_0b/\bar{v}$ moves

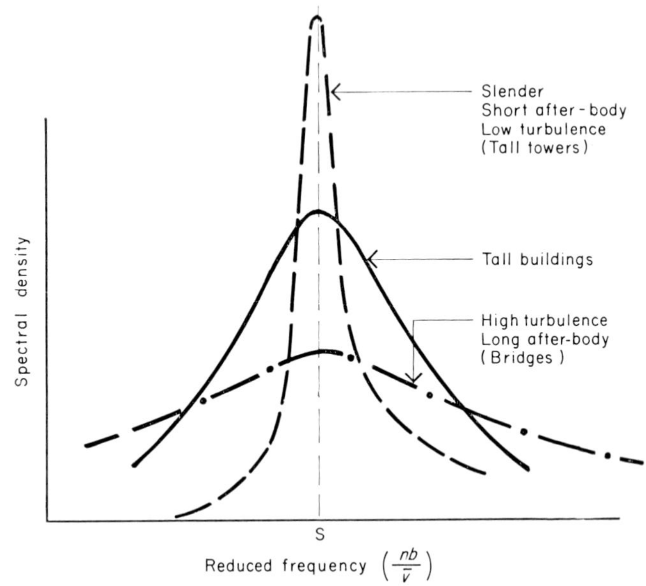

Fig. 4.21. Spectra of across-wind forces.

toward the energy containing region, the dynamic amplification becomes more significant. If the spectrum is sharp (narrow) the response will reach a maximum when $n_0b/\bar{v} \simeq S$ and will fall away at speeds above n_0b/S; the response will be essentially resonant, i.e. at the natural frequency of the structure. If the spectrum is broad there may be no maximum.

4.6.2. *Vortex Shedding*

The formation of vortices behind slender bluff bodies in smooth or low-turbulence flow is extremely regular and the force variation almost periodic; the spectrum is narrow banded and a distinct critical speed can be noted in the response.

If the structure is lightly damped and the turbulence low the amplitudes attained may be sufficient to influence the vortex formation in such a way as to force it to conform with the motion. In these circumstances, the vortex forces are periodic and at a frequency equal to the natural frequency; large and commonly destructive amplitudes will result.

Fortunately, the occurrence of severe oscillations due to vortex shedding is rare and normally restricted to lightly damped (welded steel) structures of circular or near circular cross-section located in situations of low turbulence (very tall or of moderate height in Categories 1 and 2 terrain). Methods of prediction of vortex induced motions are discussed by Vickery and Clark [4.13].

4.6.3. Broad Band Excitation

The wake spectra of tall buildings are somewhat broader than those for tall slender towers and the critical speed, if it exists at all, is commonly beyond the normal design range. Even so, the across-wind response of tall slender buildings (height/breadth > 3 to 4) is commonly greater than the down-wind or drag response. Theoretical estimates of the across-wind response are not presently feasible but semi-empirical methods based on wind tunnel tests have been suggested [4.14, 4.15]. Vickery [4.14] has suggested that the peak across-wind tip deflection of a tall building can be estimated from the equation;

$$\frac{\hat{\delta}}{(A)^{\frac{1}{2}}} = C\left(\frac{\overline{V}}{n_0(A)^{\frac{1}{2}}}\right)^n \cdot \frac{1}{\zeta^{\frac{1}{2}}} \cdot \frac{\rho}{\rho_b} \qquad (4.15)$$

where $\hat{\delta}$ = peak across-wind deflection; A = cross-sectional area; n_0 = natural frequency; ρ = air density; ρ_b = bulk density of building; $\overline{V}$ = mean speed at roof level; ζ = damping as a fraction of critical; n = 3·5 (approx.); $C = 0{\cdot}020 \pm 0{\cdot}006$.

If eqn. (4.15) suggests across-wind motions of the same order as the down-wind deformations then wind tunnel tests are strongly recommended.

4.7. Aerodynamic Damping

Turbulence and wake induced forces exist irrespective of the motion of the structure although, as noted in Section 4.6.2, vortex shedding is influenced by the motion of a structure. Aerodynamic damping forces exist only if the structure is in motion.

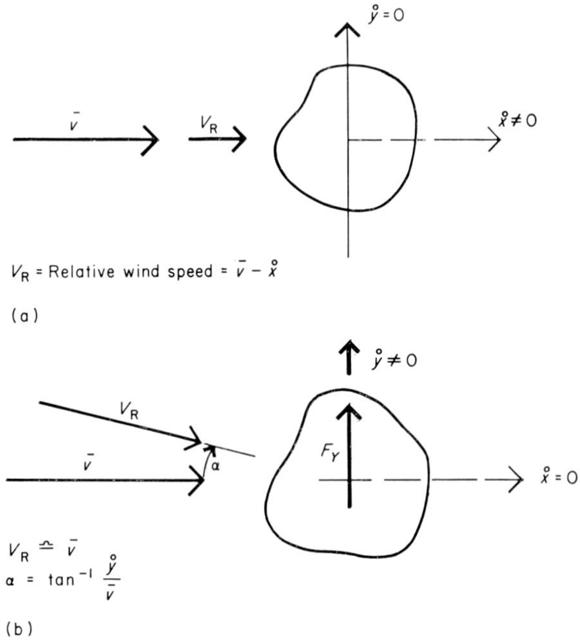

Fig. 4.22. Definition sketches for aerodynamic damping.

Figure 4.22(a) indicates the actual and relative wind speeds for a body moving at a speed $\dot{x}$ in the same direction as the mean speed $\bar{v}$. The drag force on this body is:

$$\begin{aligned} F_D &= C_D . A . \tfrac{1}{2}\rho \bar{V}_R{}^2 \\ &= C_D A \tfrac{1}{2}\rho \bar{V}^2 - C_D A \rho \bar{V}\dot{x} + C_D A \tfrac{1}{2}\rho \dot{x}^2 \end{aligned} \tag{4.16}$$

If, as is usual, $\dot{x}/\bar{V} \gg 1$ the third term of eqn. (4.16) may be ignored. The first term is the mean drag force and the second term is a viscous damping force (a force proportional to $\dot{x}$). The sign of the damping force is negative, i.e. the force opposes the motion, and energy is extracted from the body. The aerodynamic damping adds to any structural damping. For buildings the aerodynamic damping in the along-wind direction is helpful but is generally a small addition to the structural damping; only for very light structures with large frontal areas is the aerodynamic damping important. An example of such a structure might be a large outdoor movie screen.

The across-wind or lift force on a body may be expressed as:

$$F_Y = C_{Y_\alpha} . A . \tfrac{1}{2}\rho \bar{V}^2 \tag{4.17}$$

where C_{Y_α}, the lift force coefficient, is a function of the angle of attack, α, shown in Fig. 4.22(b). If the body is moving at right angles to the mean flow at a speed $\dot{y}$ which is small compared to $\dot{v}$ the lift force, F_Y, is given by:

$$F_Y = C_{Y_0} . A . \tfrac{1}{2}\rho \bar{V}^2 - \frac{\partial C_Y}{\partial \alpha} . A . \tfrac{1}{2}\rho \bar{V}\dot{y} \tag{4.18}$$

The second term of eqn. (4.18) is again a damping force. If $\partial C_Y/\partial_\alpha$ is negative this force acts in the same sense as $\dot{y}$ and energy is fed into the structure; the aerodynamic damping is negative. The magnitude of the damping force increases with $\bar{V}$ and at some speed, $\bar{V}_{crit}$, it will eventually counteract completely the positive structural damping. At this critical speed there is no damping and, as discussed in Section 4.2, the amplitude will tend to increase. For linear systems the increase is theoretically unbounded but in reality a limit is imposed by non-linear behaviour or by failure.

Aerodynamic instabilities of the type described are rare but have been observed in the behaviour of extremely slender structures such as transmission lines and guy wires. It is not a serious problem in regard to buildings but even at subcritical speeds the negative aerodynamic damping may reduce the total available damping and thus increase the sensitivity of the structure to dynamic excitation.

4.8. Torsion

Few codes draw attention to torsional wind loads although these can be significant. Perhaps the most susceptible type of a building is the tall slab-like structure which is aerodynamically susceptible to, and quite often structurally sensitive to, eccentric loads.

Figure 4.23(a) shows such a slab-like structure which is stiffened by a series of equi-spaced shear walls. The peak central deflection and the peak edge deflection at roof level are shown in Fig. 4.23(b) for a range of wind directions. In the absence of any torsion, $\hat{x}_1$ and $\hat{x}_2$ would be identical. In fact the maximum deflection at the edge is about 60% greater than the central deflection. For a structure such as this it is quite possible that a normally conservative quasi-steady

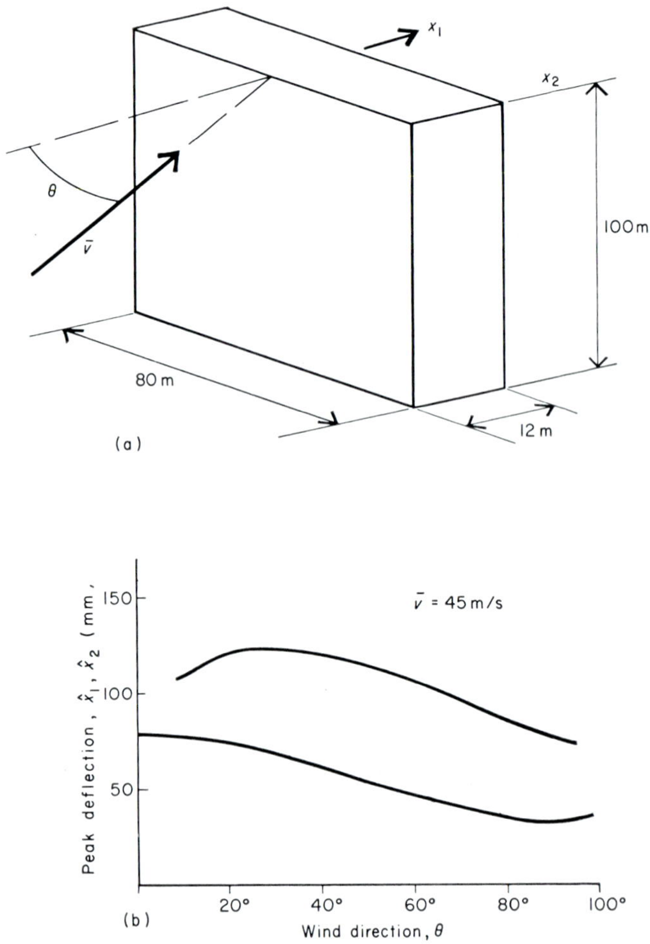

Fig. 4.23. Central and side tip deflections for a slab-like building.

'code' approach would lead to a significant underestimate of stresses in the end shear walls.

Unfortunately it is not possible to predict the magnitude of torsional loads without recourse to wind tunnel testing but care should be exercised in the case of slab-like buildings, particularly if the shear walls are equally distributed or concentrated near the centre. If the stiffness is provided primarily by end shear walls the problem is not likely to be nearly so serious.

4.9. Design Criteria

The specification of suitable criteria of acceptability with respect to wind loads and wind effects is as important as the description of the load effects themselves. The acceptance criteria discussed here include considerations of:

(i) structural failure;
(ii) cladding failure;
(iii) excessive deflections; and
(iv) discomfort of occupants due to excessive sway.

4.9.1. Structural Failure

The prime concerns of most wind codes are the strength and safety of the structure although at present this is rarely quantified in terms of an acceptable risk of failure. The risk of failure of structures designed to minimum standards is often surprisingly high (perhaps 1% or more in a 50 year life) but due to a variety of hidden safety factors the risk actually attained in practice is much lower. Davenport [4.16] suggests that for major structures the risk actually achieved could be as low as 10^{-6} in a 100 year life. The achieved failure rate for light industrial buildings is some orders of magnitude greater and certain classes of structure have quite high risks. Tall masts appear to have a failure risk of perhaps 1% in a 30 year life.

As stated, most codes avoid the specification of an acceptable risk but the question is receiving attention as more countries move toward 'limit state' design approach rather than the conventional 'working load' approach which uses design loads with a high risk of occurrence in conjunction with substantial load or safety factors.

4.9.2. Cladding Failures

Generally it should be possible to accept higher risks for the cladding than for the structural frame but this is not always so. A warehouse protecting valuable stock may have a cost well below the stock itself and the failure of the cladding might well cause a loss only marginally less than a major structural failure. More commonly, however, cladding failures are not as costly and higher risks are acceptable.

4.9.3. Deflections

Deflections under wind action are, in themselves, rarely cause for concern. Of greater significance are the effects that might accompany large deflections, e.g. cracking of finishes or 'creaking and groaning' as partition walls move relative to the main frame. As a minimum requirement, deflections under the accepted working load should not lead to significant damage to finishes.

4.9.4. Accelerations

Motion perceptible to occupants has been a cause for concern for some years. Tall buildings do move appreciably in the wind and it is

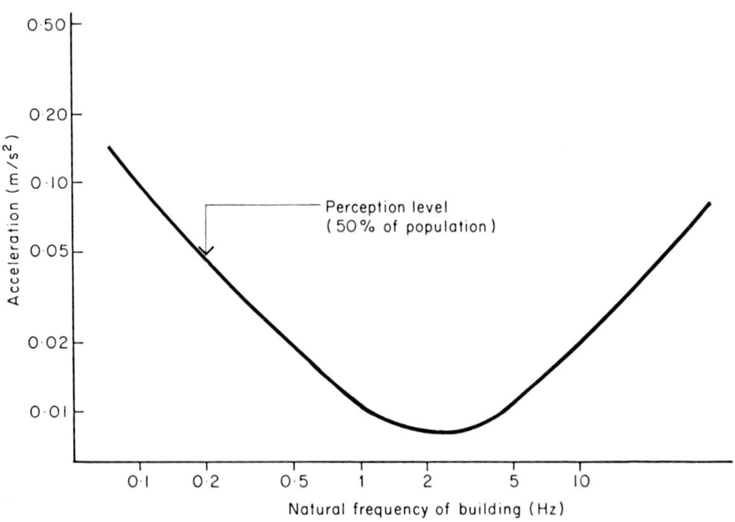

Fig. 4.24. Perception threshold for horizontal motion.

not practicable to ensure that a tall building will not move perceptibly at some time during its life. The question of acceptance involves both the level of the motion and the recurrence interval. Clearly perceptible and possibly disturbing motions can be tolerated if they occur only once or twice in the life of a structure whereas small but nevertheless perceptible motions may be unacceptable if they occur frequently.

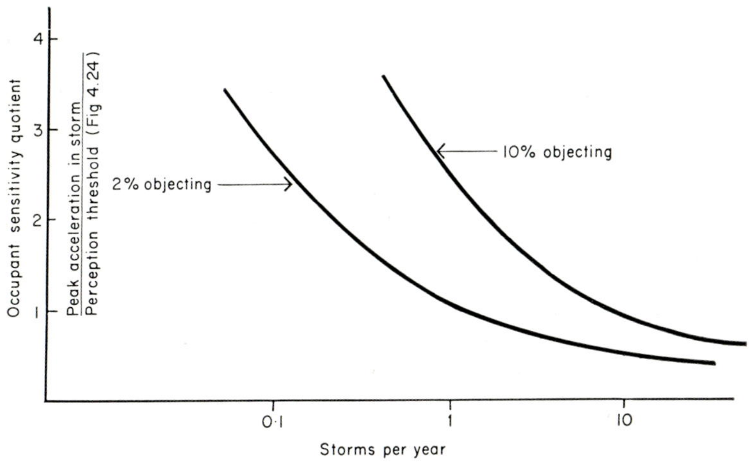

Fig. 4.25. Human sensitivity to building motion.

Acceptance criteria based on the work of Davenport [4.17], Robertson [4.18, 4.19] and Hansen *et al.* [4.20] are presented in Figs. 4.24 and 4.25. Figure 4.24 shows the acceleration perception level as a function of the frequency of vibration while Fig. 4.25 indicates acceptability levels. In Fig. 4.25 the so-called 'Occupant Sensitivity Quotient' (the ratio of the actual acceleration to the perceptible acceleration) is plotted against the recurrence interval. The curves plotted represent conditions acceptable to 2% of the population and 10% of the population. A predicted performance falling above the 10% curve is a clear indication of potential problems with wind induced motions.

Chapter 5

Environmental Wind Flows Around Buildings

The study of flow fields induced around buildings has increased in recent years mainly because a number of large buildings have been erected which, because of combinations of shape, height and isolation, have induced wind flows in public access-ways and recreational areas much higher than could be reasonably tolerated. Hence, it has become common practice now to submit building design proposals to wind tunnel testing to determine the likely wind environment around the buildings and to modify and add protection to achieve acceptable conditions. One result of these studies is an accumulation of expertise in means of avoiding high induced wind speeds. Most of this chapter will be involved with outlining flow mechanisms, good design practice for the avoidance of high induced wind speeds around buildings, the criteria involved and an example of a wind tunnel evaluation of a specific proposal.

An allied environmental problem is the dispersal of gaseous pollutants from buildings and resultant concentrations at street level and air conditioning intakes. The diffusion of pollutants is very dependent on the wind flow around buildings. Model studies of this problem to date have been relatively limited and concerned more with the location of exhaust outlets and the quantity of pollutant which can be safely discharged. An introduction to work in this area will be included in this chapter.

5.1. The Wind Flow Field Around a Building

The flow fields around bluff bodies were introduced in Section 2.1 and these will now be extended to relate to a building in the natural wind. The building is three dimensional, permitting flow around a free end, and the incident freestream flow is a turbulent boundary layer with mean velocity increasing with height and turbulence intensity decreasing with height. There are two quite separate mechanisms (pressure fields) which cause high induced wind speeds at ground level and which will be described with reference to Fig. 5.1.

The first type of flow is caused by the pressure distribution on the windward face of a building (Fig. 4.13) which is related to the local wind dynamic pressure which increases with height. The resulting pressure gradient, decreasing with height, induces flow vertically

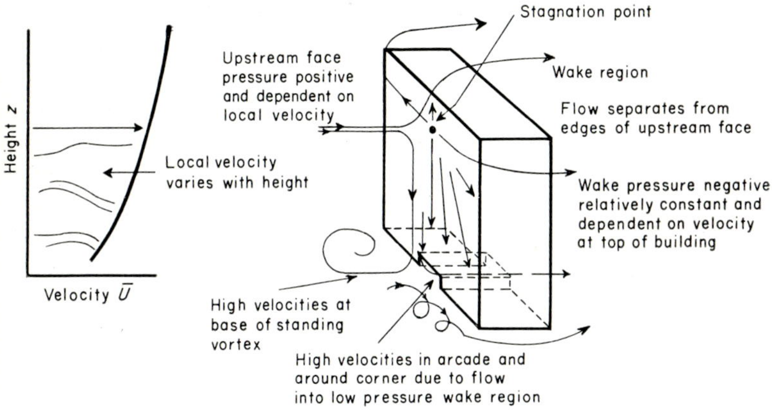

Fig. 5.1. Flow field around a building (after Melbourne [5.1]).

down the face below the stagnation point (which can be as high as 80% of the building height). This flow rolls up into a standing vortex system at the base of the building (Fig. 5.2), causing high wind speeds in this region. Buildings of near circular platform which promote lateral flow, do not produce strong vertical flows. Conversely, rectangular and concave buildings do produce strong vertical flows with consequent high wind conditions in the standing vortex system. The configuration of upstream buildings can be critical for this flow because under certain conditions the vortex flow behind a lower upstream building can augment the vortex in front of the larger building, further increasing the high wind speeds at the base as shown in Fig. 5.3.

The second type of flow is caused by the pressure difference between the low pressure wake regions (leeward and side faces) and the relatively high pressure regions at the base on the windward face.

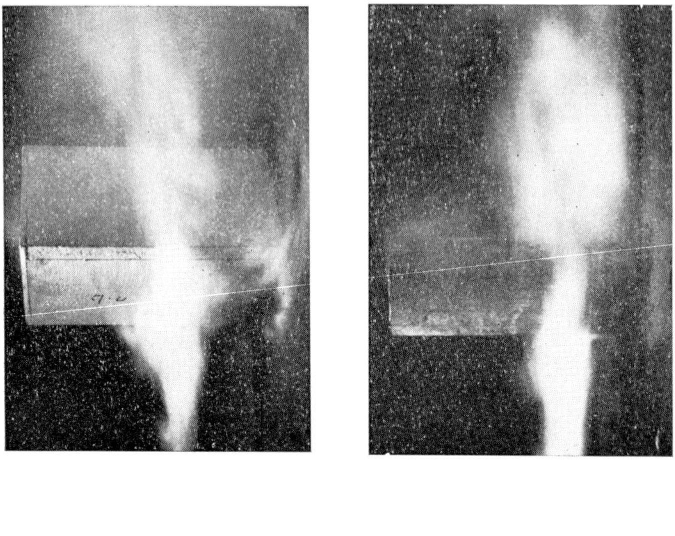

Fig. 5.2. Visualisation of flow around a rectangular building.

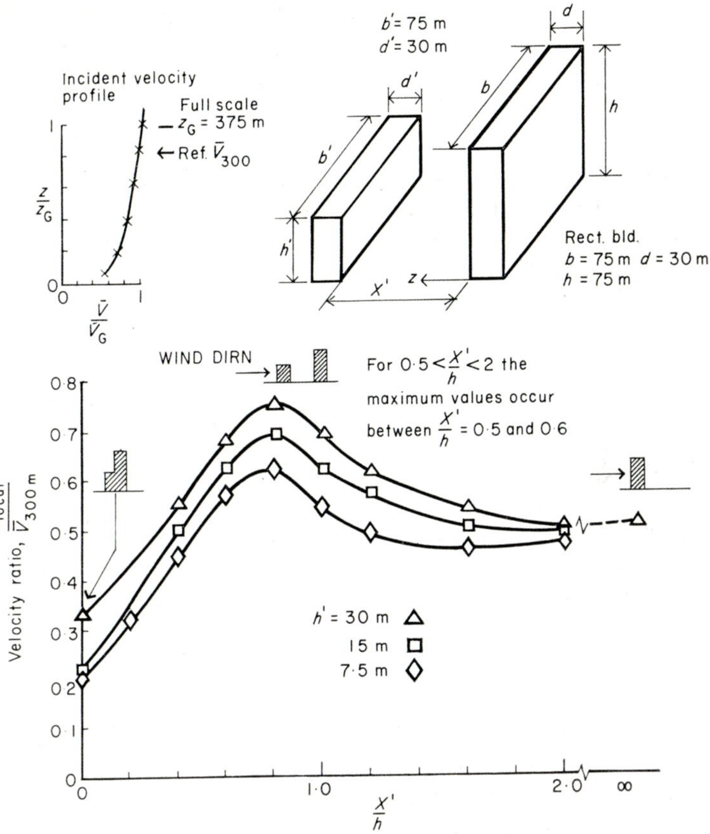

Fig. 5.3. Ground level wind speed measurements in front of a large building being affected by a lower upstream building (after Melbourne [5.1]).

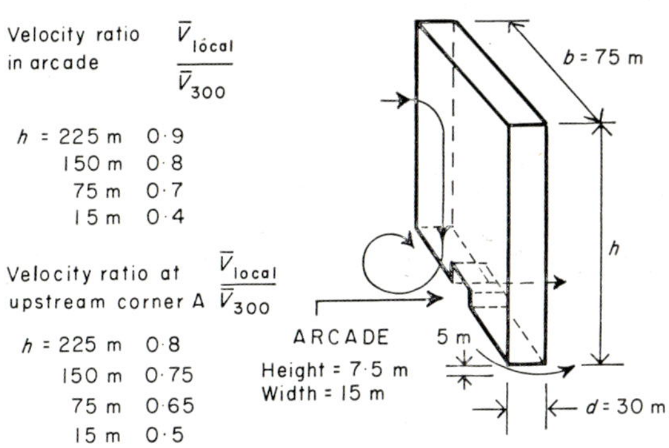

Fig. 5.4. Typical ground level wind speeds induced in an arcade and near the corner of a building as a function of building height (after Melbourne [5.1]).

Flow directly between these two regions through arcades or around corners can cause very high local wind speeds. The low wake pressure is dependent on the velocity along the top free boundary, that is, the freestream velocity at the top of the building, as described in Section 2.3. Hence the taller the building the lower the wake pressure and the higher the velocities which are induced through arcades and around corners for a given aspect ratio, as shown in Fig. 5.4. In general, this problem is much harder to control because the wake pressure cannot easily be modified.

5.2. Environmental Wind Speed Criteria

Before discussing techniques for predicting, avoiding or ameliorating environmental wind conditions, it is necessary to establish some sort of criteria as to the acceptability or unacceptability of certain wind speeds. That is to form some assessment of the effects of wind on people as they use main pedestrian access-ways near buildings, recreational areas or just plain window shopping.

The wind force felt by a person is related to dynamic pressure. Hence, whilst it may be convenient in one sense to relate criteria directly to wind speed, it must be appreciated that the force felt by a person is proportional to wind speed squared. That is, double the wind speed results in four times the force. For this reason a more rational feel for the problem is gained if comparative data is presented in terms of velocity pressures rather than velocities. However, criteria referred to wind speed have gained popular acceptance and values of wind speed are more easily remembered than numbers based on the square of wind speed. Hence, criteria will be discussed in terms of wind speed but comparisons will be given using velocity and velocity pressure approaches.

5.2.1. Wind Conditions which Cause People to be Blown Over

Melbourne [5.1a] described conditions which caused people to be blown over in the forecourt of the Menzies Building at Monash University as follows:

> 'The wind was gusting regularly up to 20 m/s which caused all people in the area great difficulty with balance, several staggering

a number of paces to regain balance, particularly when the gust caught them sideways. The people concerned in the area were all relatively young and few were carrying anything more than a brief-case. On separate occasions two girls were unbalanced to the extent that they came down onto their hands and knees on the adjacent lawn. The maximum gust velocity recorded during this period was 23 m/s rising from about 12 m/s in 2–3 s. It is concluded from these observations that wind gusts of 20 m/s will unbalance and seriously inconvenience a number of people and wind gusts of 23 m/s and over will cause people to be blown over. Obviously a person can stand in a much higher steady wind velocity but it appears to be the unexpected nature of the peak gust which catches most people off guard and gives rise to the above observations.'

Experience since this report has given little reason to alter this assessment that at gust speeds of 23 m/s, particularly from the side, people will commence to be unbalanced to the point of being brought to their knees. A reasonable extension of this assessment would be to conclude that wind conditions with maximum wind gust speeds in excess of 23 m/s could be regarded as quite dangerous to any pedestrian movement.

5.2.2. *Acceptable Wind Conditions*

At the other end of the scale the assessment of what wind conditions are acceptable is more difficult because it is much more subjective and dependent on the activity concerned and the temperature and humidity of the environment concerned. Melbourne [5.1b] restricted his assessment to what was acceptable in main public access-ways and suggested that maximum wind speeds per annum in the old city environment in Australian cities, before the advent of high rise buildings, were about 16 m/s. That is a gust speed producing about half of the pressure of that which caused people to be blown over ($16^2 \approx \frac{1}{2} 23^2$). He found that these 'old city' wind conditions in public access-ways were generally regarded as acceptable and so this was suggested as a starting point for defining acceptable wind conditions.

The architect is often more concerned with the creation of areas which people wish to frequent because they are environmentally attractive. These areas range from plaza restaurants, through window shopping precincts to parks which require a further subjective assessment with respect to wind environment beyond the criteria so far discussed with respect to public access-ways. Davenport [5.2b]

Table 5.1

Tentative comfort criteria for windiness (after Davenport [5.2*a*])
(Units: Beaufort number; temperatures >10°C.)

Activity	Areas applicable	Relative comfort: Perceptible	Tolerable	Unpleasant	Dangerous
1. Walking fast	Sidewalks	5	6	7	8
2. Strolling, skating	Parks, entrances, skating rinks	4	5	6	8
3. Standing, sitting —short exposure	Parks, plaza areas	3	4	5	8
4. Standing, sitting —long exposure	Outdoor restaurants, bandshells, theatres	2	3	4	8
Representative criteria for acceptability			<1 occn/ week	<1 occn/ month	<1 occn/ year

At lower temperatures relative comfort level might be expected to be reduced by one Beaufort number for every 20°C reduction in temperature.

Relationship between Beaufort number and wind speed (m/s)

Beaufort number	1	2	3	4	5	6	7	8
Mean wind speed in range at 10 m height in open country	0·9	2·4	4·4	6·7	9·3	12·4	15·5	18·9
Gust speed at 2 m height from $\hat{V}_2 = \bar{V}_{10}\ 1{\cdot}5 \times 0{\cdot}8$	1	3	5	8	11	15	19	23
Mean wind speed in range at 2 m height in city area from $\bar{V}_2 = 0{\cdot}5\ \hat{V}_2$	0·5	1·5	2·5	4	5·5	7·5	9·5	11·5

developed the above work of Melbourne [5.1b] and that of Wise [5.3] further by considering various activities and areas and suggested a table based on the Beaufort scale and which is reproduced as Table 5.1. Care should be taken in interpreting the wind speeds given by Davenport in Table 5.1 as they are mean wind speeds and these were related to the Beaufort scale referenced to ocean or open country conditions at a height of 10 m where the turbulence intensity is relatively low and the conversion from gust wind speed to mean wind speed used was

$$\hat{V} = 1{\cdot}5\,\overline{V}$$

The equivalent gust velocities at 2 m height on this basis have been

Table 5.2

Summary of wind effects on people based on the Beaufort scale (after Penwarden [5.6])

	Beaufort number	Speed (m/s)	Effects
Calm, light air	0, 1	0–1·5	Calm; no noticeable wind
Light breeze	2	1·6–3·3	Wind felt on face
Gentle breeze	3	3·4–5·4	Wind extends light flag Hair is disturbed Clothing flaps
Moderate breeze	4	5·5–7·9	Raises dust, dry soil and loose paper Hair disarranged
Fresh breeze	5	8·0–10·7	Force of wind felt on body Drifting snow becomes airborne Limit of agreeable wind on land
Strong breeze	6	10·8–13·8	Umbrellas used with difficulty Hair blown straight Difficult to walk steadily Wind noise on ears unpleasant Windborne snow above head height (blizzard)
Near gale	7	13·9–17·1	Inconvenience felt when walking
Gale	8	17·2–20·7	Generally impedes progress Great difficulty with balance in gusts
Strong gale	9	20·8–24·4	People blown over by gusts

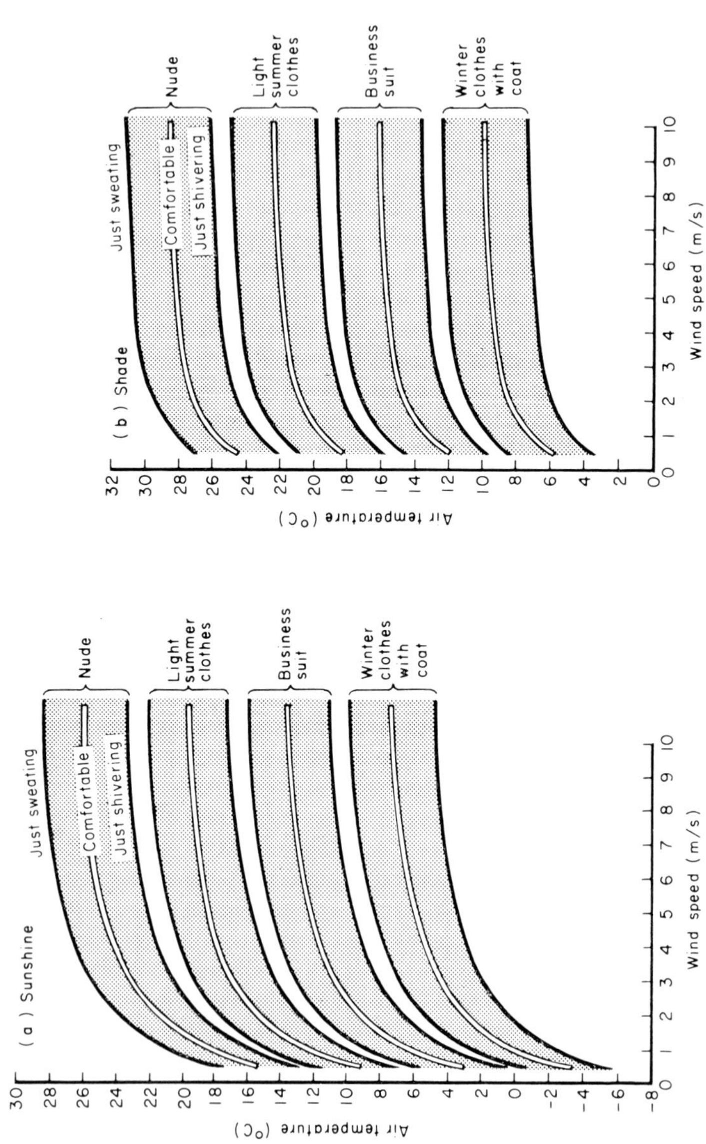

Fig. 5.5. Examples of comfort conditions for strolling (after Penwarden [5.6]).

added to Table 5.1 (where the correction for height used was $\hat{V}_2 = 0{\cdot}8\ \hat{V}_{10}$). As will be discussed in Section 5.2.3, city situations have higher turbulence intensities where frequently the relationship is more like

$$\hat{V} = 2{\cdot}0\ \bar{V}$$

and values for $\bar{V}$ on this basis have also been added to Table 5.1.

Penwarden [5.6] has also produced a summary of wind effects on people along the lines of a Beaufort Scale, but with peak gust wind speeds, which has been reproduced in Table 5.2, and he also developed two figures to describe comfort conditions for strolling in sunshine and in shade which have been reproduced as Fig. 5.5.

5.2.3. *Relationships Between Peak Gust and Mean Wind Speeds and Frequency of Occurrence*

The turbulent properties of the natural wind boundary layer, in particular the probability distribution of wind speed, has been introduced in Section 3.2.1. It was noted that the peak gust is dependent on turbulence intensity and values recorded by most meteorological anemometers (response time 1–3 s) can be given in terms of the mean and standard deviation as

$$\hat{V} = \bar{V} + 3{\cdot}5\ \sigma_V \tag{5.1}$$

For example, for a turbulence intensity ($\sigma_V/\bar{V}$) of 15%, $\hat{V} = 1{\cdot}5\ \bar{V}$, and for 30%, $\hat{V} = 2{\cdot}0\ \bar{V}$, etc.

It is worth commenting further on the relationship between the peak gust and mean wind speeds because it has a very important bearing on the prediction of wind environment via wind tunnel testing to be discussed in the next section. It seems reasonable to conclude that it is the peak gust wind speeds and associated gradients which people feel most and as such it is of interest to know under what conditions they occur. The observations of Melbourne and Joubert [5.1] indicated that the areas in full scale which have been classed as having unpleasant or unacceptably high wind speeds were all associated with high mean wind speeds.

Later, model and full scale measurements by Isyumov and Davenport [5.4b] and Melbourne [5.5] continued to show that the windiest areas were associated with high mean wind speeds, but that the turbulence intensity was important in determining the peak gust wind speeds. In the case of the former, the ratio of peak gust wind speed over mean wind speed $\hat{V}/\overline{V}$ for the three windiest conditions respectively was 1·5, 2·7 and 2·8 and for the latter 1·9, 1·9 and 2·4; for areas and wind directions with lower wind conditions, and obviously for much greater turbulence intensities, this ratio was typically as high as 5·0. This means that to get an accurate prediction of peak gust wind speeds from wind tunnel model tests, it is essential that mean and r.m.s. or peak values for a given probability level be actually measured.

Although it is possible to have unpleasant areas with low mean wind speeds and high turbulence intensities, the evidence to date does seem to indicate that areas likely to have unacceptably high wind conditions, such as near corners, in narrow alleys and in arcades, the turbulence intensity is relatively low and that in these areas it would be reasonable to assume that the peak gust wind speeds will be about twice the mean wind speed. This means that wind tunnel investigations in terms of establishing and improving likely areas of unacceptably high wind conditions can be reasonably based on very simple and inexpensive model measurements of mean wind speed. However, this does not mean that the need to model the turbulence characteristics of the incident wind stream can be overlooked, as a low turbulence stream would produce quite different flow fields and erroneous information as discussed in Section 2.10 and by Isyumov and Davenport [5.4a].

It is possible to look at the environmental wind criteria over a full probability range against which model measurements, properly integrated with wind data, can be compared. In Fig. 5.6, Melbourne's criteria are shown developed in terms of cumulative probability distributions for daylight time per annum. Similar distributions can be prepared for various seasons and different time blocks in the day if required. In the lower part of the figure the distribution of the maximum gust speeds per annum, of 23 and 16 m/s, are shown as a normal distribution back to the maximum hourly mean wind speed per annum, for $\hat{V} = 2{\cdot}0\ \overline{V}$ and $\sigma_V = 0{\cdot}3\ \overline{V}$. In the upper part of the figure the cumulative distribution of hourly mean wind speeds has been shown as a Weibull distribution with $K = 2$, that is a Rayleigh distribution.

As well as a knowledge of the total number of hours per annum

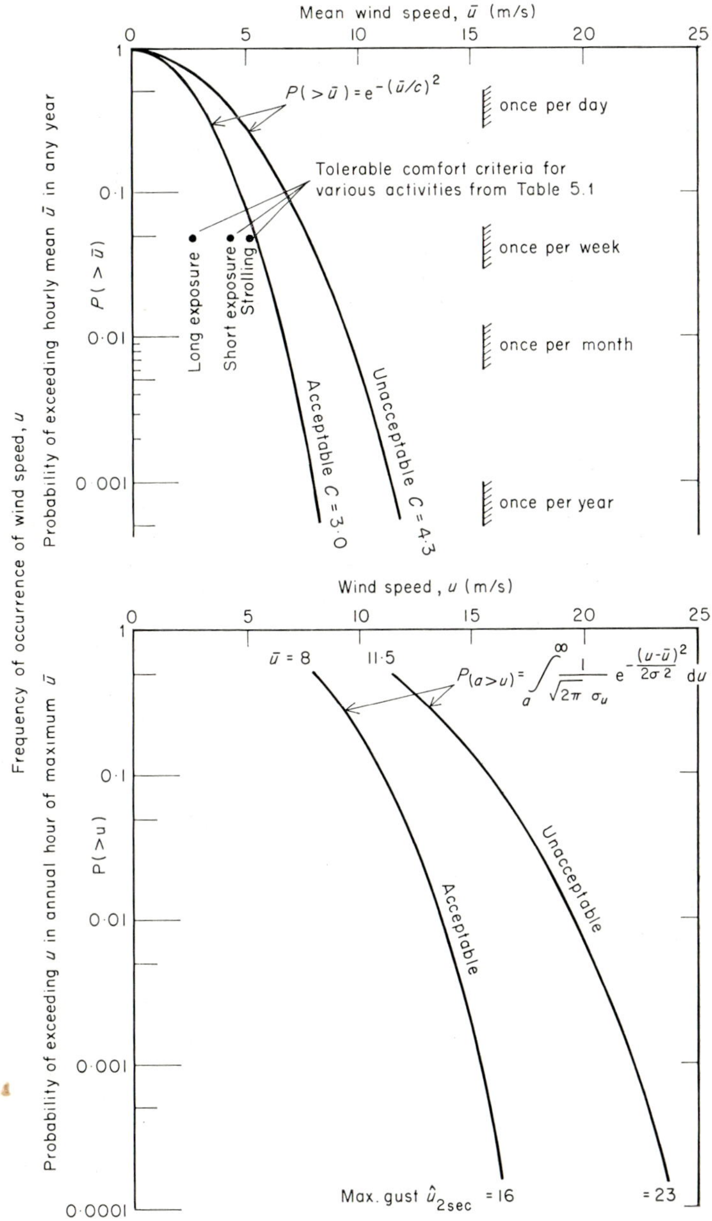

Fig. 5.6. Cumulative probability distributions of Melbourne's Criteria for ground level wind speeds in public access -ways,for $\hat{V} = 2{\cdot}0\ \bar{V}$ and $\sigma_V = 0{\cdot}3\ \bar{V}$, in terms of daylight time per annum for which a given wind speed is exceeded, and onto which Davenport's Comfort Criteria has been superimposed.

that a mean wind speed is exceeded (i.e. $P_{(>V)}$) it is useful to know the number of storms per annum during which a mean wind speed is exceeded, or alternatively the magnitude of storms with average frequencies of occurrence of 1/annum, 1/month, 1/week and 1/day. Davenport [5.4a] has given a method for doing this and these average frequencies of occurrence have been added to Fig. 5.6. This then allows a superimposition on Fig. 5.6 of Davenport's Comfort Criteria (Table 5.1), for the four activities and relative comfort which have been related to frequencies of occurrence of less than 1/week, 1/month and 1/year respectively.

In relation to the criteria presented in Fig. 5.6, it is interesting to note a procedure used by the U.K. Building Research Establishment in determining likely 'User Satisfaction' based on full scale experience as reported in Ref. 5.7a. They have found that if the frequency of exceeding a mean wind speed of 5 m/s exceeds 20% (876 daylight hours per annum) protective measures are likely to be needed; between 10% and 20% complaint is likely but may be insufficient to provoke action, and below 10% (438 daylight hours per annum) conditions are likely to be satisfactory.

Whilst the environmental criteria displayed in Fig. 5.6 are aimed at covering the majority of real situations, it must be remembered that this has been prepared for situations where the turbulence intensity is 30% ($\sigma_V = 0{\cdot}3\,\overline{V}$). If the turbulence intensity in some cases was found to be significantly above or below 30% (or $\hat{V} \neq 2\,\overline{V}$) then an adjustment to the values shown in Fig. 5.6 should be made. In practice, however, if the criteria are based on the premise that it is the peak gust wind speed and associated gradients which people feel most and the turbulence intensity at a given location changes significantly with wind direction, then one would only be justified in making decisions based on the frequency of occurrence of peak gust wind speeds. For the more sophisticated model measuring techniques, where peak values for a given probability level are recorded, this is the more rigorous approach and in some ways is more simple to use.

Bearing in mind the subjective nature of the decisions made, the prediction accuracy required does not really warrant great sophistication. In fact, simplicity and speed are the main factors required in practice.

5.3. The Prediction of Wind Environment Around a Building

Although some experience has been gained from assessing full scale wind conditions about a variety of buildings, it is necessary to turn to wind tunnel model testing if a confident prediction of the wind environment induced around a specific building proposal is required. In a model test the wind speeds at a number of locations can be measured for different wind directions and combined with local wind data to permit the frequency of occurrence of wind speeds at those locations to be calculated and compared with environmental wind speed criteria.

5.3.1. Wind Tunnel Testing Techniques

To model the turbulent flow conditions induced by buildings, it is necessary to place the model building in a model of the natural wind, which as a minimum must reproduce the correct mean velocity and turbulence intensity profiles as described in Section 2.11.2.

A model of the specific building complex along with adjacent buildings is required in sufficient detail to ensure that the aerodynamic flow fields are satisfactorily modelled. In general, surface features up to a metre in depth do not need to be modelled, as they have little effect on the flow field around a building; however, details of this size can be significant locally, that is at ground level, and should be included. In addition to the specific building complex and adjacent buildings, it is essential to model all major buildings for at least 800 m radius as the shielding effect of single buildings can be very significant. It is not usually possible to mount these additional buildings on a wind tunnel turntable and it is normal practice to have them just as single block models which can be placed upstream for the relevant wind directions. An example of such a model set up is shown in Fig. 5.7.

There are varying degrees of sophistication in the techniques which can now be used to measure the effect of wind on people at the model scale. As previously discussed, a person feels the wind as a force or pressure distribution which varies as the square of the wind speed. The ideal way to measure this effect then is to measure the forces on a model person. Unfortunately at most model scales this is a very small target and although it has been attempted, it is a very difficult instrumentation problem and cannot be generally recommended. There are two simple alternatives, either to measure pressure with a total head and static tube, or velocity with a hot wire (or hot

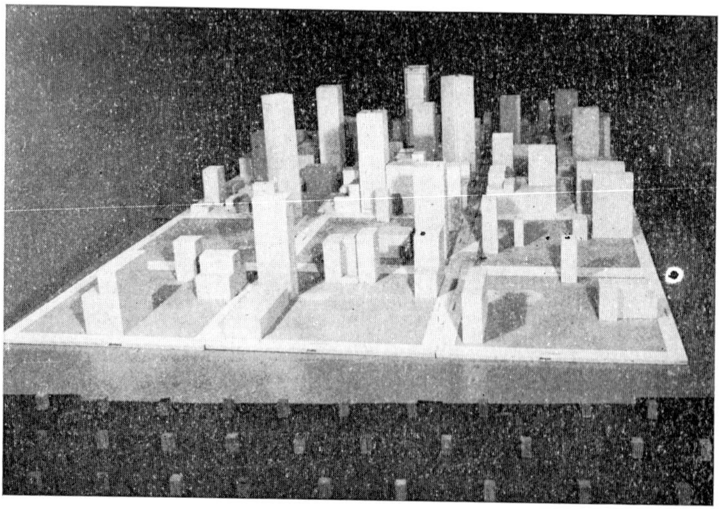

Fig. 5.7. A 1/400 scale wind tunnel model set up for the measurement of environmental wind conditions in surrounding access-ways.

film) probe. In both cases the velocity pressure or velocity measured has to be referenced to a velocity pressure or velocity measured in the freestream at some height or distance which is free from the influence of the buildings. It is convenient to make this reference at or near gradient height or alternatively at a known anemometer site. The main aim is to have the reference at a point where the meteorological data of wind speed frequency of occurrence can be established, which then allows frequency of occurrence of wind speeds at each station under test to be established by virtue of a simple ratio.

The velocity pressure and hot wire velocity measuring techniques both have advantages and disadvantages. The hot wire technique has problems in that hot wire measurements in turbulence intensities over 20% become increasingly suspect. In particular, the hot wire is insensitive to wind direction in the plane normal to the wire, hence reverse flows appear as a positive addition to the mean velocity measured by this technique. As most environmental wind flows involve turbulence intensities above 20% and frequently above 30%, the mean velocity measured by a hot wire anemometer is quite unreliable. However, if only peak velocities are used, in particular those obtained from a probability analysis for a fixed low probability level, then the problem is to a large extent overcome because there is no dependence on a mean velocity measurement. On the advantage side the hot wire, used with the wire vertical, records velocity independent of wind direction; hence, once the probe has been placed at a measuring station, the model can be rotated for different wind directions without any further adjustment of the probe. For the same reason there is the disadvantage in not gaining some knowledge of the local flow direction. Hot wire and associated equipment is relatively expensive.

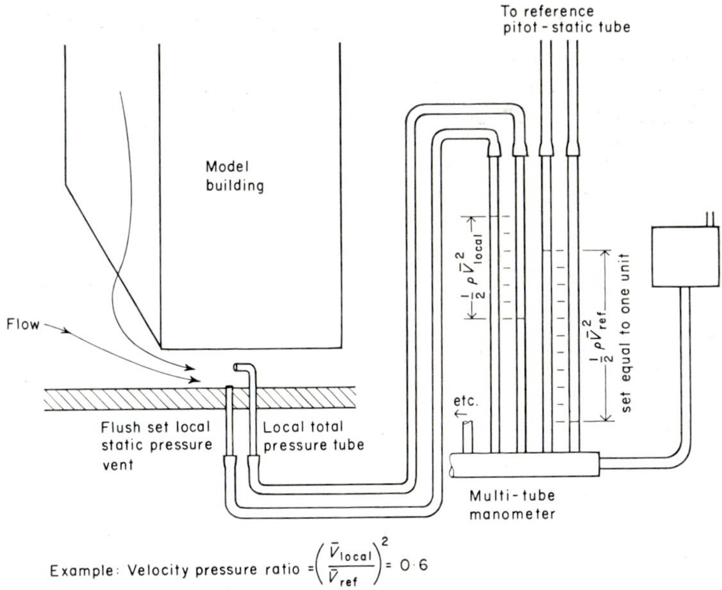

Fig. 5.8. Set up for measuring mean velocity pressure ratios in a model environmental wind flow investigation.

The velocity pressure can be simply measured by using a length of small diameter tube (e.g. 1–2 mm diameter), bent in the horizontal plane to measure total pressure and a surface static vent can be provided with a similar tube as shown in Fig. 5.8. If only mean velocity pressure (dynamic pressures) are being measured, then the local static and total pressures from a number of stations can be displayed on a multitube manometer and the difference, which is the dynamic pressure, can be compared directly as a ratio with the reference dynamic pressure as shown in Fig. 5.8. This method requires very little expensive equipment and measurements at a number of stations can be made at the same time. The disadvantage is that the total pressure tubes have to be adjusted to point into the local flow direction for every change of total wind direction. The further disadvantage is that to get fluctuating or peak values, a pressure transducer and associated electronic equipment have to be used which involves the same degree of expense as the hot wire equipment.

5.3.2. Example of a Wind Tunnel Test

Probably the best way to describe the total process of obtaining a prediction of wind environment around a building from a model test is to follow through a complete example.

This example will be of the simplest type of test where only mean measurements are made and the comparisons with criteria are made on the assumption that for most cases $\hat{V} = 2\,\bar{V}$ as given in Fig. 5.6. A hypothetical configuration of buildings will be used, with the main investigation being centred on a rectangular building of height 183 m (600 ft), and sides 46 m (150 ft) and 30 m (100 ft). These are the dimensions of the building being promoted by the Commonwealth Aeronautical Advisory Research Council (CAARC) for use as a Standard Tall Building Model for the Comparison of Simulated Natural Winds in Wind Tunnels, Wardlaw and Moss [5.9]. The plan layout of buildings to be used for this example, given in Fig. 5.9, has no relation to any known configuration of buildings.

Before commencing the wind tunnel test, it is advisable to take account of the regional wind speed data and to assemble environmental wind criteria in a form which can be used directly during the testing programme. A basic wind speed probability distribution can be obtained from balloon measurements or ground station measurements and then referred to a gradient height or some other reference by the methods given in Chapter 3. If a full probabilistic presentation of the predicted wind environment is required, the wind tunnel data has eventually to be combined with the wind speed probability data and one of the most convenient ways of doing this is to assemble the latter in a series of Weibull distributions for $22\frac{1}{2}$ degree wind direction

Fig. 5.9. Plan of CAARC Building and surrounding buildings for example of environmental wind tunnel test.

segments as described by Isyumov and Davenport [5.4b]. However, for many investigations, it has been found convenient, and sufficient for most decision making exercises, to adopt local criteria based on annual maximum events which have the implied distributions

Table 5.3

Environmental wind criteria based on average annual maximum events for the CAARC Building example having the implied distributions given in Fig. 5.6

Wind direction	Average annual hourly maximum wind speed at 300 m for each sector	Velocity ratio, $\bar{u}_{\text{local}}/\bar{u}_{300}$ to give $\bar{u}_{\text{local}} = 11{\cdot}5$ m/s (i.e. $\hat{u} \simeq 23$ m/s)	Velocity pressure ratio $(\bar{u}_{\text{local}}/\bar{u}_{300})^2$ to give $\bar{u}_{\text{local}} = 11{\cdot}5$ m/s (i.e. $\hat{u} \simeq 23$ m/s)
N	21	0·55	0·30
NNE	19	0·61	0·37
NE	11	>1	>1
ENE	6	>1	>1
E	6	>1	>1
ESE	10	>1	>1
SE	13	0·88	0·78
SSE	13	0·88	0·78
S	17	0·68	0·46
SSW	17	0·68	0·46
SW	18	0·64	0·41
WSW	19	0·61	0·37
W	19	0·61	0·37
WNW	18	0·64	0·41
NW	18	0·64	0·41
NNW	19	0·61	0·37

given in Fig. 5.6. For this purpose it is necessary to compile a table of average annual hourly maximum wind speeds at the reference position. From these, criteria contours, in terms of the ratio of the local area reference measurement can be obtained as illustrated in Table 5.3 and Fig. 5.10. The criteria are given here as velocity ratios and velocity pressure ratios, to illustrate the relative insensitivity of the velocity ratio when it is velocity squared effects which should be taken into account in any subjective decision making processes involved during the test programme.

The test of the environmental wind conditions induced around the

CAARC Building example might proceed along the following lines. The measurements given in the example are of velocity pressure ratios obtained by the method illustrated in Fig. 5.7; a similar description would apply to direct velocity ratio measuring techniques.

If the CAARC Building was placed on the block with vertical walls rising directly above a colonnade the layout and induced wind

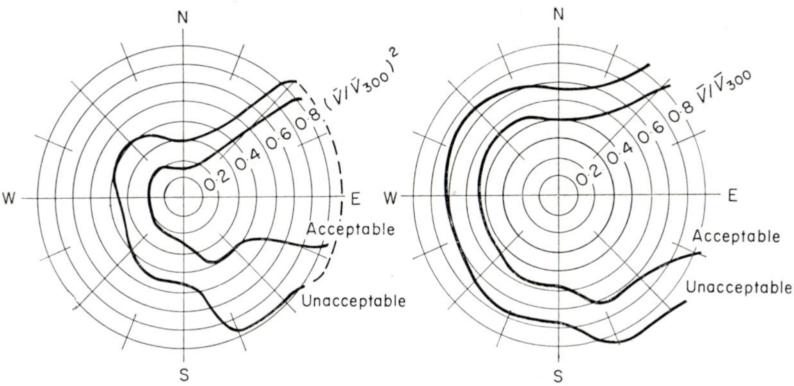

Fig. 5.10. Contours of environmental wind criteria from Table 5.3.

flows would be as shown in Fig. 5.11 for Configuration 1. After the manner described in Section 5.1 the building in this configuration would induce a large amount of vertical flow down the exposed north face which would roll up into a standing vortex at the base and spill laterally around the colonnade and into the side street. The magnitudes of the velocity pressure ratios induced by this configuration are given in Fig. 5.11 and can now be compared directly with the criteria given in Fig. 5.10. Conditions at the NE corner in the colonnade (Station A) and even some distance away (Station B) would give rise to wind speeds which would cause people to be blown over on several occasions per annum and as such would be unacceptable. In particular, the velocity pressure ratios induced on the opposite side of the street at Station C would be quite unacceptable because of the northerly wind direction for which they occur. As one other piece of absurdity the velocity pressure ratios induced in an open arcade, Station D, are given. Open arcades in this type of situation only have

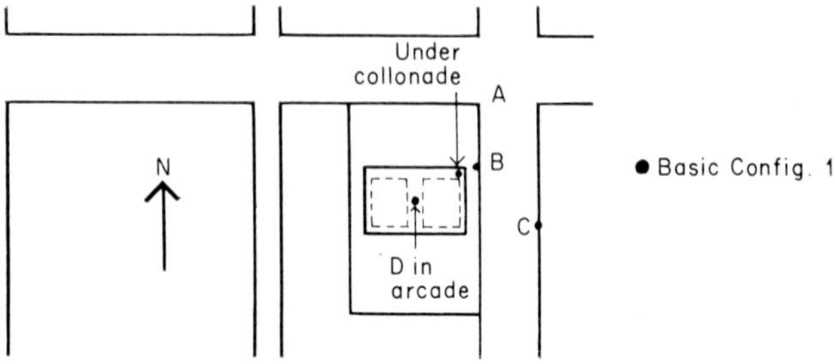

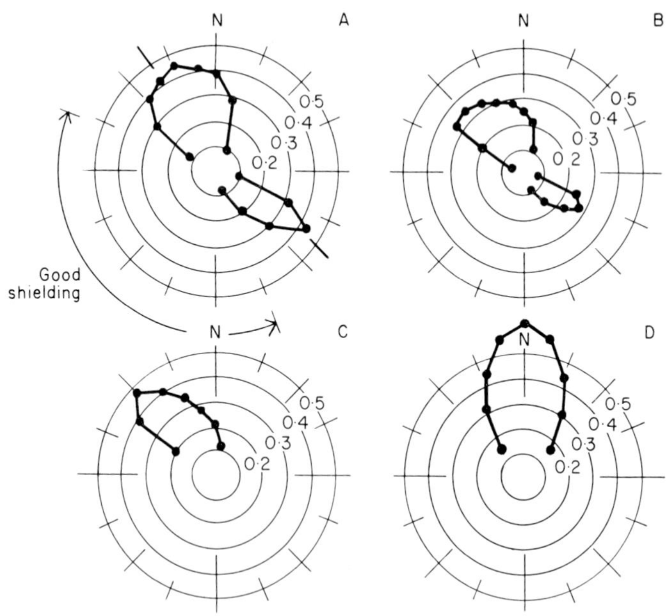

Fig. 5.11. CAARC Building—Configuration 1 Velocity Pressure ratio $(\bar{V}_{local}/\bar{V}_{300\ m})^2$.

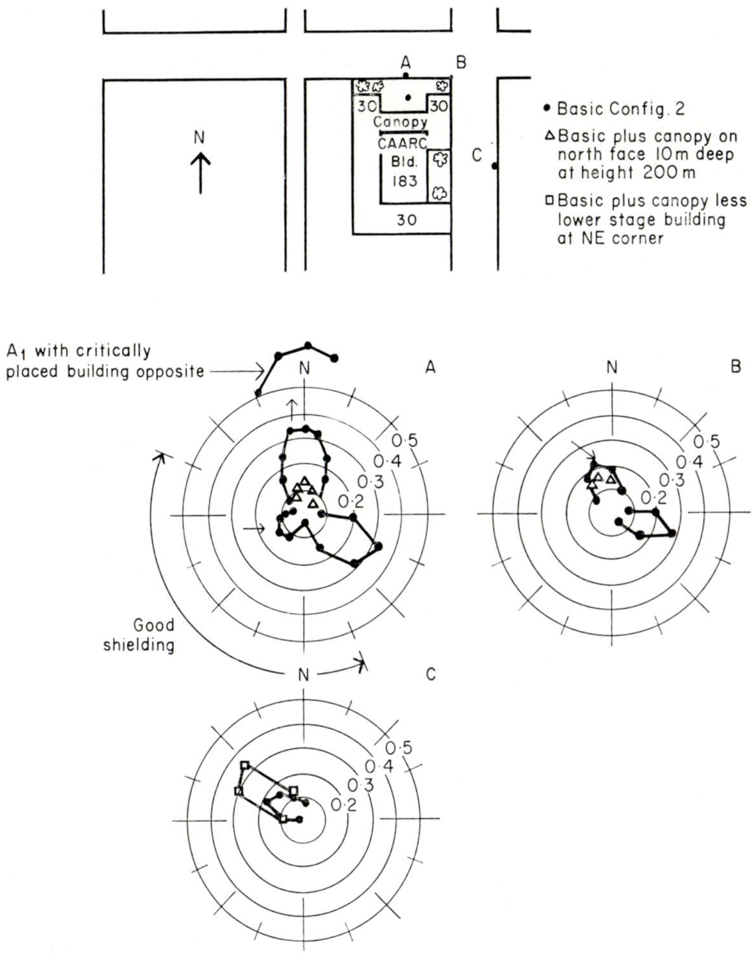

Fig. 5.12. CAARC Building—Configuration 2 Velocity Pressure ratios $(\bar{V}_{local}/\bar{V}_{300\ m})^2$.

a hope of being acceptable if there are large buildings immediately behind (in the south in this case) which can cause a back pressure at the leeward entrance to the arcade. Normally arcades in such a building must be sealed.

Virtually the only way to achieve satisfactory conditions in an open forecourt to the north, and prevent excessive flow into the surrounding streets and retain the rectangular building shape is to make extensive use of lower stage buildings and canopies as set out in Fig. 5.12 as Configuration 2. For the critical Stations A, B and C, it is shown that this treatment will bring induced wind flows down to the acceptable level defined in Fig. 5.10. However, the addition of the canopy is essential to lift the standing vortex clear of the forecourt. In particular the potentially disastrous effects of having a building of height about $\frac{1}{3}h$ opposite at the critical distance of $0{\cdot}8\,h$, which augments the standing vortex, is also shown. The effect of the low

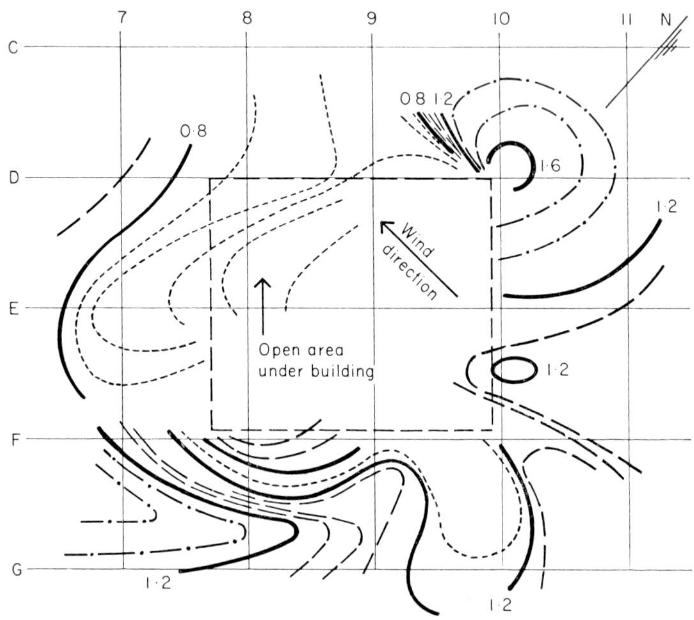

Fig. 5.13. Map of ground level velocity ratios in the vicinity of a building (after Vickery [5.10]).

stage building at the NE corner in preventing flow reaching the side street, Station C, is clearly shown when it is removed with velocity pressure ratios going to well in excess of 0·3 even with the canopy which has little effect on flow this far away from the tower.

There are a number of more sophisticated forms of presentation of model data varying from complete probability distributions such as could be overplotted on Fig. 5.6, for each station to a wind map of a complete area. An example of a wind map, after Vickery [5.10] is shown in Fig. 5.13; here the contours shown are velocity ratios of local ground level gust speed over mean wind speed at an anemometer site for which long term records were available. The gust speed is the 95% speed, i.e. the speed exceeded for about 5% of the time during which mean speed at the reference anemometer is constant. Maps of this type, obtained for a number of wind directions, can be integrated with the anemometer records to yield a total picture of the ground wind climate.

5.4. Guide Lines for Avoiding High Induced Wind Speeds at Ground Level

With a knowledge of the wind flow fields induced around buildings, it is possible for the architect to create designs which avoid causing high induced wind speeds at ground level. The following notes have been set down to provide some guidance in this respect.

(1) Buildings will only induce high ground level wind speeds if a significant part of the building is exposed to direct wind flows. It is in fact the direct exposure of a building face to high speed wind flows rather than building height alone which causes the problem. For example, a 20 m high building at an airport site can induce higher wind speeds at a corner than a 100 m high building in a city centre situation. If a building has to be built on an exposed site, then much more attention to some of the following guidelines is needed.

(2) In what situation can a proposed building be regarded as being either exposed or adequately shielded? A general rule is that a building should be regarded as potentially exposed

for any particular direction if half of the building height is clearly above upstream buildings which can provide shielding. An upstream building of half height is reckoned to shed a wake region of low wind speeds which will cover about three quarters of the downstream building under consideration, and the direct wind flow striking the building above this wake will tend to be deflected up and around rather than downwards. The shielding effect of upstream buildings is one of the most significant factors to be taken into account when an assessment of environmental wind conditions is being made and the shielding provided by single buildings can be effective for between 500 and 1000 m downstream.

(3) If a building is regarded as being exposed for some or all wind directions, then there are some fundamental configuration factors which can be taken into account.

(a) Rectangular buildings with a main face rising clearly from ground level, normal to any of the unshielded prevailing wind directions, should be avoided as these are the most common cause of environmental wind problems.

(b) Rectangular buildings placed on a podium or lower stage building complex which deflects the downward wind flow before it reaches ground level, can be satisfactorily used in many cases. Canopies (preferably with parapets) can also be used to deflect this flow and lift a standing vortex above pedestrian level. However, canopies (unless very extensive) do little to ameliorate high wind speeds induced at the corners of buildings or in colonnades.

(c) Buildings with circular or near circular planform promote lateral flow and induce very little downward vertical flow and hence are to be preferred in exposed situations. Circular buildings do induce high wind speeds at the maximum width normal to the wind direction, similar to that induced at the corners of rectangular buildings. However, in the case of circular buildings, this is caused by acceleration of the local flow around the circle which can be reduced to a large extent by relatively low height upstream shielding; this is not so for rectangular buildings, because much of the flow around the corners originates from flow induced down the windward face from higher levels.

(4) Arcades and colonnades at the base of exposed buildings can provide openings between the high upstream face pressures and low base pressures through which high wind speeds can be induced. In the case of arcades, it is usually necessary to provide means of sealing them with revolving doors or mechanically operated sliding doors on strong wind days. Colonnades should not be designed as main public accessways or as window shopping precincts unless adequate upstream shielding is assured.

(5) There seems to be no reason why high wind speeds cannot be accepted in certain areas, such as near corners, in narrow passages between buildings or on tops of podiums, provided such areas do not have to be used for general public access. For example, gardens and architectural features can be placed at a corner to prevent pedestrian access close to the corner and to provide a local wind break for areas immediately downstream of the corner. An open podium area set up for recreational purposes would only be used when weather conditions (including wind) are pleasant. Narrow passages between buildings are often only used for vehicular access and as such can tolerate much higher wind conditions.

(6) The architect is mostly concerned with the wind environment in the immediate vicinity of the particular building being designed. In most cases, even when buildings are exposed, it is usually possible to provide canopies, local wind breaks, sealed arcades, etc., to achieve satisfactory wind conditions in adjacent areas (even after the event). However, it is most important to realise that one of the worst features of exposed rectangular buildings is the increased wind flow induced at ground level generally, and particularly on the opposite side of the street as shown in Fig. 5.11. This type of problem is by far the most difficult to deal with, because much of it is caused by pressure and flow fields well away from the face of the building concerned and consequently local changes within the

site boundaries have little effect. To overcome this problem it is often necessary to undertake major configuration changes, such as reducing height, changing planform or setting a tower well back on a podium. As such, it is obvious that this type of problem cannot be solved easily after a building has been constructed, and must be taken into account in the very early design stages.

5.5. Dispersal of Gaseous Pollutants

The complete field of the dispersal and diffusion of gaseous pollutants in the atmosphere is outside the scope of this text, but some mention can be made of work in this area which has a dependence on wind flow around buildings. In this category there are two types of gaseous pollutant problems which are currently of interest to architects.

(1) The height of a chimney stack, carrying boiler or other types of gaseous exhaust, must be sufficient to ensure that the exhaust plume clears the immediate buildings, otherwise there is no point in having a chimney stack.

(2) Where pollutants are exhausted into wake regions of buildings, it is necessary to determine the resultant concentrations which occur at air conditioning intakes and at street level.

In recent years both of these problems have been the subject of model testing on an *ad hoc* basis. These problems depend so much on the specific configurations involved that it is difficult to give any generalised guidance. Even the old rule of thumb that to be really effective a stack should be twice as high as adjacent buildings cannot be quoted as a certain way of avoiding problems.

The phenomenon to be studied is essentially that of a stream of buoyant contaminating fluid entering either a boundary layer stream or a wake region. The path and spread of the contaminating plume in the initial phases are dependent primarily on exit buoyancy and momentum conditions in relation to the local freestream momentum as described by Melbourne [5.11]. To achieve dynamic similarity for

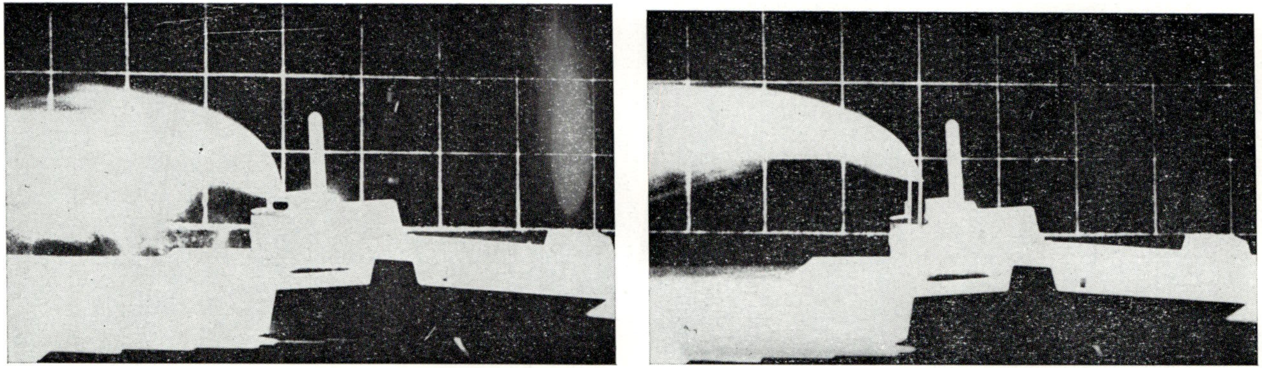

Fig. 5.14. Example of model tests to determine the height and location of a chimney stack to avoid local entrainment of the plume in the building wake.

model studies, it is necessary to keep the following non-dimensional scaling parameters equal in both scales.

Plume buoyancy $\dfrac{(\Delta\rho)gD}{\rho V^2}$

Initial plume momentum $\dfrac{\rho_p W^2}{\rho V^2}$

Plume mass flow rate $\dfrac{\rho_p W D^2}{\rho V L^2}$

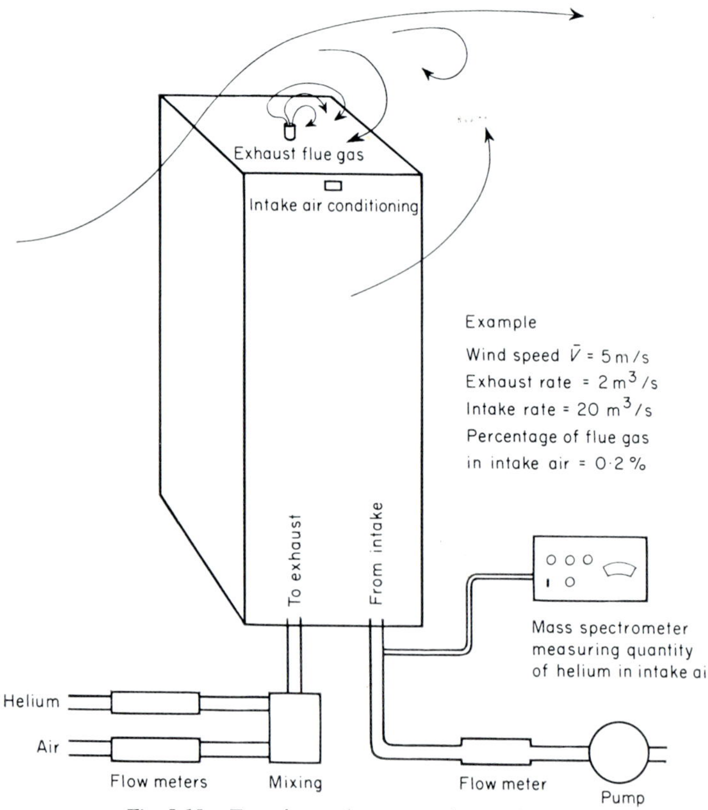

Fig. 5.15. Experimental set up and several results.

where V, L and ρ are freestream velocity at building height, building width dimension and freestream density, respectively, and D, W, $\Delta\rho$ and ρ_p are initial plume diameter, vertical velocity, density defect $(\rho - \rho_p)$ and density respectively, and g is the gravitational acceleration.

It is usual to use helium as the buoyant exhaust gas in model studies to achieve as high a model velocity as possible (even so model freestream velocities are mostly of the order of 1 m/s).

In Fig. 5.14 an example is given of the way in which a plume path can be influenced by the flow field around a building. The plume path is traced visibly by introducing smoke (vaporised oil) into the exhaust gas. For given wind conditions it is then possible to determine a suitable stack location and height to avoid direct entrainment of the plume into the building wake or premature ground strike of part of the plume.

A description of a model study of the contamination of air conditioning intakes by flue gas exhausted into the wake region at the top of a tall building has been given by Melbourne and Gartshore [5.12]. This type of study involves the measurement of pollutant concentrations and one of the most satisfactory ways to do this is to use a helium plume, or a plume with a helium tracer, and measure the very low levels of helium concentrations at various locations with a mass spectrometer tuned to helium. An example of this type of measurement is given in Fig. 5.15.

Chapter 6

Natural Ventilation

Despite the availability of mechanical ventilation and air conditioning equipment, many buildings, due to cost limitations, rely on natural ventilation to achieve minimum ventilation for health or, air movement to improve indoor comfort in warm humid climates. It must be pointed out that there are many occasions during which natural ventilation will not occur and temporary use of fans is necessary if ventilation or air movement is to be maintained.

6.1. Ventilation Rates for Carbon Dioxide and Odour Control

Rates of ventilation required by building regulations or suggested by professional groups such as ASHRAE or ISHVE vary widely depending on the criteria adopted. Minimum ventilation rates are those associated with maintaining the proportions of metabolic gases, oxygen and carbon dioxide, in occupied spaces.

It was found by Consolazio and Pecora [6.1] in 1947 that 1·7 m^3 of fresh air per person per hour is adequate for this purpose in

Function of space	Ventilation rate (m^3/h fresh air)
Residences	
Living room	45
Kitchen	60–90
Bedroom	25
Bathroom	25
School classrooms	40 per person
Factories (no smoking)	30 per person
Community halls (heavy smoking)	50 per person

Note: As ventilation requirements vary widely it is wise to check on local practice before proceeding with designs for natural ventilation.

Fig. 6.1. Typical ventilation rates for carbon dioxide and odour control.

submarines. Criteria used for ventilation in building spaces in temperate climates are usually related to odour reduction, in particular body odour, cooking smells and tobacco smoke. Ventilation rates to reduce odours to an acceptable level vary from 25 to 90 m^3 per hour depending on the volume and function of the space. In large spaces where the number of occupants vary, ventilation rates are given per person and calculated on the maximum number of occupants. Typical minimum ventilation rates for spaces which are often naturally ventilated are indicated in Fig. 6.1.

6.2. Ventilation to Improve Thermal Comfort in Hot Humid Environments

Ventilation is often used to improve the thermal comfort of building occupants in hot humid environments. In this application ventilation criteria do not depend directly on volumetric air change rates, but more on local air speeds in occupied portions of the space. The cooling effect of the airflow on the occupants is due to convection and evaporation of sweat from the skin. The evaporative capacity of the air decreases with increases in relative humidity.

A number of complex indices have been developed to describe the influences of climatic and physiological factors on human comfort. These indices have been summarised and compared by Givoni [6.2]. Considering the vagaries of natural wind and the subjectivity of what constitutes 'comfort', the order of accuracy provided by a simple method developed by Macfarlane [6.3] is probably adequate when considering natural ventilation.

Combining data from his own studies with those by Yaglou and Ellis, Macfarlane [6.3] suggested a method of adjusting dry bulb temperature comfort zones for variations in latitude, relative humidity, radiant heat and air movement. The reference comfort zones were given as:

Location	Summer comfort zone dry bulb temp. (°C)	Winter comfort zone dry bulb temp. (°C)
Greater than 30° lat.	21–28	20–25
Less than 30° lat.	24–30	21–28

Variations in comfort zones allow for acclimatisation and differences in clothing. Correction to comfortable summer dry bulb temperature zones are:

(1) Decrease comfortable dry bulb temperatures by 0·8°C for each 10% increase in relative humidity above 60%.
(2) For radiant heat subtract 0·55°C for each 2·8°C increase in radiating surfaces above 38°C.
(3) For air movement add 0·55°C for each 0·15 m/s up to 1·0 m/s for dry bulb temperatures below 37°C.

Good building practice, in hot humid climates, includes provision for shading or thermal insulation in the building envelope to eliminate high internal surface temperatures as sources of radiant heat gain to occupants.

Assuming such a practice is followed, corrections to Macfarlane's comfort zones can be reduced to those for relative humidity and air movement. On this basis, air speeds necessary to restore comfort in environments with high dry bulb temperatures and relative humidities can be determined from Fig. 6.2.

6.3. Energy Sources for Natural Ventilation

There are two energy sources that create air pressure differences used to promote natural ventilation:

(1) Pressure differences due to variations in air density with height, created by differences in air temperature, referred to as the 'stack effect'.
(2) Pressure differences due to distribution of wind pressures on buildings.

Pressure differences from stack effect and wind pressure distribution can act simultaneously, the effective pressure being their algebraic sum. However, when wind is present the pressure difference due to wind is generally dominant.

Application of the stack effect to ventilation is limited to situations

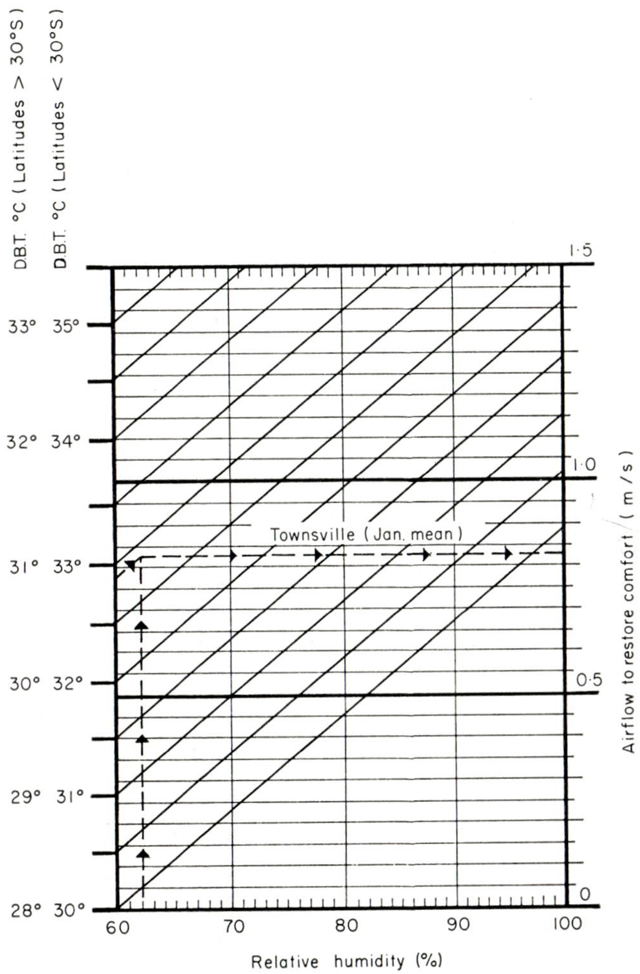

Fig. 6.2. Minimum airflow to restore comfort for a range of dry bulb temperatures and relative humidities.

where there are significant differences between internal and external air temperatures and differences in height between inlet and outlet openings in the building envelope. Favourable conditions for ventilation utilising the stack effect, consistent with thermal comfort of building occupants, occur in multistorey buildings in regions with cold winter climate. Heating of building interiors for winter comfort creates a significant indoor–outdoor temperature difference which can be used to provide minimum ventilation rates by the stack effect. Multistorey buildings increase the height differences between wall openings for ventilation and reduce the area of external wall and roof surfaces exposed to low outdoor temperatures.

Buildings in warm humid climates often use natural ventilation due to wind to improve the thermal comfort of occupants as well as to supply fresh air for building spaces. Elongated buildings, orientated normal to prevailing winds, increase wall area available for openings and decrease the resistance to airflow created by small openings in internal partition walls. Extended eaves and projecting end walls on the windward faces encourage even distribution of wind pressures over the windward face which improves cross ventilation through rooms near the ends of the buildings.

Buildings in temperate climates often utilise stack effect in winter and wind pressures in summer. As climatic extremes are less in these regions, building forms are more varied with less emphasis on optimising for summer airflow or winter heat loss.

6.4. Applications of Stack Effect for Natural Ventilation

In cold climates with low external temperatures and heated building interiors temperature differences can be large, of the order of 30°C. This potential for natural ventilation is sometimes used to provide minimum ventilation by infiltration through small cracks around doors and windows, to prevent carbon dioxide and odour levels reaching uncomfortable levels [6.4].

Stack effect in buildings in hot climates is usually insignificant as openings are larger and ventilation due to wind ensures temperature differences between internal and external air temperatures are small, of the order of 2°C. Exceptions can occur in spaces where high temperature processes are being performed such as foundries, steelworks or commercial kitchens but airflow due to stack effect in such cases is unlikely to offset the high heat stress on workers.

Equation (2.28) in Chapter 2 for discharge due to stack effect assumes that inlet and outlet openings are equal in area. Where this is not the case, a correction factor should be used, determined from Fig. 6.3 for the appropriate opening area ratio.

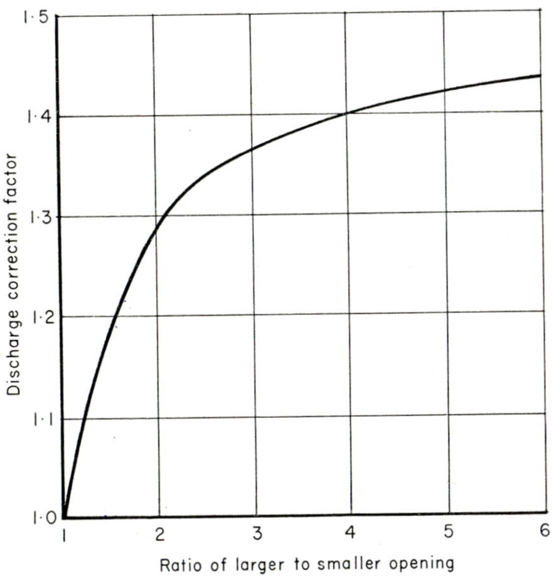

Fig. 6.3. Correction factors for relative areas of inlets and outlets.

Airflow efficiency of an opening in a building is accommodated in the coefficient C. For most openings in buildings a coefficient of 0·12 is recommended. Equation (2.28) in Chapter 2 can then be written:

$$Q = 0{\cdot}12\, A(H(T_2 - T_1))^{\frac{1}{2}} \text{ m}^3/\text{s} \tag{6.1}$$

or

$$Q = 420\, A(H(T_2 - T_1))^{\frac{1}{2}} \text{ m}^3/\text{h} \tag{6.2}$$

6.4.1. Example

A small two storey residence (Fig. 6.4) is heated during winter to

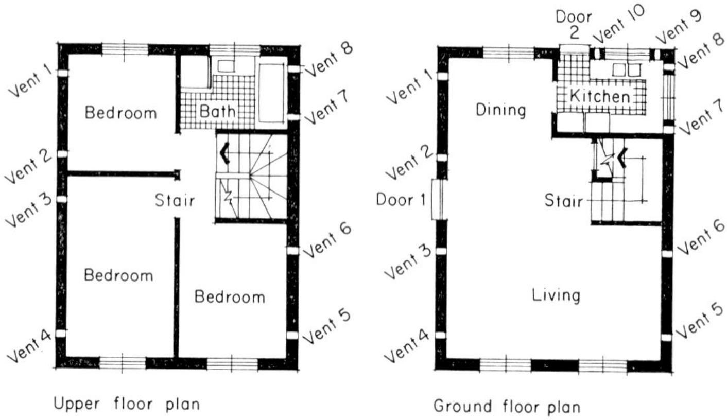

Fig. 6.4.

an average indoor air temperature of 25°C when the outdoor temperature is −5°C. Estimate the ventilation due to stack effect using the data provided.

Data

Free area of openings in upper floor. 8 wood casement window sashes without weatherstripping. Perimeter of each frame 3 m. Free area of crack per metre length of crack 605 mm^2 [6.5]

Total free area = 0·015 m^2

8 fixed vents each with a free area of 1000 mm^2

Total free area = 0·008 m^2

Total free area of openings upper floor = 0·023 m^2

Free area of openings in lower floor. 10 wood casement window sashes without weatherstripping. Perimeter of each window 3 m. Free area of crack per metre length of crack 605 mm^2.

Total free area = 0·018 m^2

10 fixed vents each with a free area of 1000 mm^2

Total free area = 0·010 m^2

2 timber doors without weatherstripping each with free crack area of 3000 mm^2

Total free area = 0·012 m^2

Total free area of openings lower floor = 0·040 m^2

Average difference in height between openings on upper and lower floors is 3 m.
Coefficient for all openings is 420

Average indoor air temperature $= 25+273 = 298$ K
Average outdoor air temperature $= -5+273 = 268$ K

Calculations

$$
\begin{aligned}
Q &= 420\,A(H(T_2-T_1))^{\frac{1}{2}} \\
&= 420 \times 0{\cdot}023\,(3(298-268))^{\frac{1}{2}} \\
&= 91{\cdot}6 \text{ m}^3/\text{h}
\end{aligned}
$$

Ratio of lower floor openings to upper floor openings

$$
\begin{aligned}
&= \frac{0{\cdot}040}{0{\cdot}023} \\
&= 1{\cdot}7
\end{aligned}
$$

From Fig. 6.3 correction factor for discharge is 1·23.
Corrected discharge is then

$$
\begin{aligned}
&= 91{\cdot}6 \times 1{\cdot}23 \\
&= 112{\cdot}7 \text{ m}^3/\text{h}
\end{aligned}
$$

As the free area of openings is equal for all four rooms on the upper floor the ventilation rate for each room is:

$$\frac{112{\cdot}7}{4} = 28{\cdot}2 \text{ m}^3/\text{h}$$

As the minimum ventilation rates for bedrooms and bathrooms indicated in Fig. 6.1 is 25 m^3/h, ventilation is satisfactory on the upper floor.

Distribution of airflow through the kitchen and living room on the lower floor will be proportional to the free area of openings in each room.

$$
\begin{aligned}
\text{Ventilation in living room} &= \frac{0{\cdot}023}{0{\cdot}040} \times 112{\cdot}7 \\
&= 64{\cdot}8 \text{ m}^3/\text{h}
\end{aligned}
$$

Minimum ventilation rate from Fig. 6.1 is 45 m^3/h indicating ventilation is satisfactory.

$$\text{Ventilation in kitchen} = \frac{0{\cdot}017}{0{\cdot}040} \times 112{\cdot}7$$

$$= 47{\cdot}9 \text{ m}^3/\text{h}$$

Minimum ventilation rate from Fig. 6.1 is 60 m^3/h indicating the above ventilation rate is inadequate.

This situation could easily be corrected by opening a window during periods when the kitchen was in use. Permanent ventilation at 60 m^3/h would lead to unnecessary increases in heating costs, as use of domestic kitchens is intermittent.

6.5. Applications of Wind Forces to Natural Ventilation

Natural ventilation, utilising pressure differences due to wind, has its widest application in residential, school and industrial buildings during summer in temperate and hot humid climates. Texts on tropical architecture by Koenigsberger, Fry, Givoni and Olgyay [6.6, 6.7, 6.2 and 6.8] all contain discussion of natural ventilation by wind. Emphasis was placed on natural ventilation of school buildings by Van Straaten [6.9] and Caudill and Reed at the Texas Engineering Experiment Station [6.10]. Studies related to industrial buildings are described in publications by Baturin [6.11] and Weston [6.12]. Natural ventilation of office buildings by both stack and wind effects was described by Jackman [6.4].

To achieve maximum airflow through a building, its form, orientation and exposure must be such that the pressure differences between locations of openings are maximised with respect to local wind characteristics. With the pressure difference established, the airflow through a series of openings is largely dependent on the discharge coefficient and area of the smallest opening.

Theoretically, air will flow into or out of a building with one small opening exposed to wind pressure until equilibrium is reached between internal and external air pressures in accordance with the Gas Law (eqn. (2.26)) in Chapter 2.

The theoretical equilibrium described above rarely occurs in the case of large typical door or window openings due to variations in wind pressure across such openings. Significant pressure variations occur across openings inclined to the wind, particularly those with projecting reveals. These pressure differences allow air to enter and exhaust through the same opening, sustaining airflow through the opening which fluctuates with wind gusting.

Extensive model studies by Givoni [6.13] indicated that average internal air speeds 1·2 m above floor level inside rooms with a single floor to ceiling opening 1/3 of width of the windward wall, were low, ranging from 13% to 14% of the external wind speed 1·2 m above ground, depending on wind direction. Similar models with two openings in the same wall, each 1/6th of the wall width, had average internal air speeds from 12% to 21% of external wind speed depending on wind direction. Similar cross ventilated models, with openings 1/3 of the width of the wall in windward and leeward walls, had average internal air speeds from 35% to 42% of external wind speed. As the main application of natural ventilation of buildings is for evaporative cooling of occupants during hot humid conditions, emphasis in this chapter will be placed on the higher internal air speeds offered by cross ventilation.

Two approaches can be used to estimate airflow through buildings due to wind. One approach makes use of wind pressure differences across a building, and size and airflow discharge characteristics of building openings in a discharge equation, to estimate volumetric flow rate. The other approach involves wind tunnel tests of models to determine the ratios of wind speeds at points of interest inside the building to an external reference wind speed at a height for which wind records are available. Both methods require a separate evaluation for each of the prevailing wind directions at a site.

6.5.1. Airflow Through an Opening Due to a Static Pressure difference

Where winds create a static pressure difference between opposite faces of a wall or roof, airflow through small openings in such a wall can be derived using energy eqn. (2.20) outlined in Chapter 2.

Such pressures are usually in the form of pressure coefficients and for convenience the height for the reference dynamic pressure should be chosen to match the anemometer height, usually 10 m, from which reliable wind records are available. Where there are differences between reference dynamic pressure height and anemometer height, a correction factor determined from the appropriate mean velocity profiles should be used.

6.5.1.1. Example

Pressure coefficients from wind tunnel model study based on reference dynamic pressure of approaching wind at ridge height (Fig. 6.5) need correction if local wind records to be used are measured at the standard height of 10 m above ground.

If p_1 is the wind pressure at a particular point on the surface of the model then C_{p_1} using a 4 m reference dynamic pressure is:

$$C_{p_1} = \frac{p_1}{\frac{1}{2}\rho v_4{}^2}$$

A pressure coefficient C_{p_2} for the same point using a 10 m reference dynamic pressure is:

$$C_{p_2} = \frac{p_1}{\frac{1}{2}\rho v_{10}{}^2}$$

$$= C_{p_1}\left(\frac{v_4}{v_{10}}\right)^2$$

From the mean velocity profile (Fig. 6.5)

$$v_4 = v_G\left(\frac{z_4}{z_G}\right)^{\alpha}$$

and

$$v_{10} = v_G\left(\frac{z_{10}}{z_G}\right)^{\alpha}$$

so that

$$\frac{v_4}{v_{10}} = \left(\frac{z_4}{z_{10}}\right)^{\alpha}$$

$$= \left(\frac{4}{10}\right)^{0.25}$$

$$= 0{\cdot}795$$

and

$$\left(\frac{v_4}{v_{10}}\right)^2 = 0{\cdot}63$$

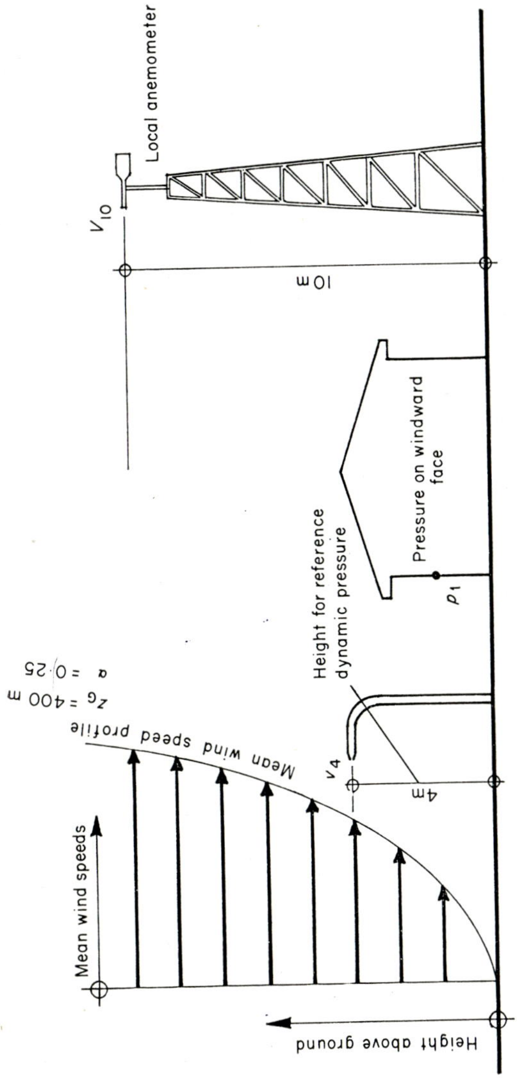
Local anemometer
V_{10}
10 m
Pressure on windward face
p_1
Height for reference dynamic pressure
v_4
4 m
Mean wind speed profile
z_G = 400 m
α = 0·25
Mean wind speeds
Height above ground

Fig. 6.5.

All pressure coefficients based on 4 m reference dynamic pressure should be multiplied by 0·63, if used directly with design wind speeds from local 10 m anemometer wind records.

By expressing the windward and leeward pressures p_1 and p_2 in eqn. (2.20) in terms of pressure coefficients and the pressure loss through a single opening p_l in terms of dynamic loss coefficient C as in eqn. (2.18) the equation becomes:

$$\tfrac{1}{2}\rho v_z^{\,2}(C_{p_1}-C_{p_2}) = C\tfrac{1}{2}\rho v_0^{\,2} \tag{6.3}$$

where v_0 is the mean air speed through the opening in the building and v_z is the reference wind speed at a height z used to determine pressure coefficients C_{p_1} and C_{p_2}.

Equation (6.3) can be rearranged so that

$$v_0^{\,2} = \frac{1}{C}(C_{p_1}-C_{p_2})v_z^{\,2}$$

or

$$v_0 = \frac{1}{(C)^{\frac{1}{2}}}((C_{p_1}-C_{p_2})v_z^{\,2})^{\frac{1}{2}} \tag{6.4}$$

Researchers studying natural ventilation have tended to use the discharge coefficient C_d rather than dynamic loss coefficient C. These coefficients are related so that

$$C_d = \frac{1}{(C)^{\frac{1}{2}}} \tag{6.5}$$

Typical values for both coefficients for a range of opening geometries are provided in A.S.H.R.A.E. Guide [6.14]. Using discharge coefficients eqn. (6.4) becomes:

$$v_0 = C_d(C_{p_1}-C_{p_2})v_z^{\,2})^{\frac{1}{2}} \tag{6.6}$$

or in terms of volumetric flow rate

$$Q = C_d A((C_{p_1}-C_{p_2})v_z^{\,2})^{\frac{1}{2}} \tag{6.7}$$

6.5.2. *Airflow Through Openings in Series*

Most applications of natural ventilation due to wind pressure differences are related to cross ventilation with air entering and leaving a building through openings in opposite or adjacent walls. As the airflow through a building passes through two or more

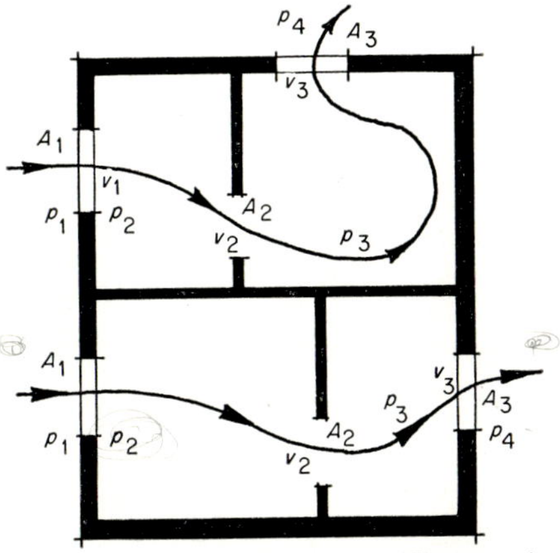

Fig. 6.6. Typical airflow paths through building openings in series.

openings (Fig. 6.6) it is convenient to develop an equation for flow through openings in series, particularly as most data on pressure distributions is related to external building surfaces.

Consider the volumetric discharge Q through three openings having areas A_1, A_2 and A_3 and discharge coefficients C_{d_1}, C_{d_2} and C_{d_3} with pressure differences $p_1 - p_2$, $p_2 - p_3$ and $p_3 - p_4$. The total pressure difference across the series of openings is:

$$(p_1 - p_2) + (p_2 - p_3) + (p_3 - p_4) = (p_1 - p_4)$$

From the principle of continuity expressed in eqn. (2.3) in Chapter 2, the volumetric rate of discharge is equal for each of the openings in series so that from eqn. (6.7):

$$Q = C_{d_1}A_1((C_{p_1} - C_{p_2})V_z^{\,2})^{\frac{1}{2}} = C_{d_2}A_2((C_{p_2} - C_{p_3})v_z^{\,2})^{\frac{1}{2}}$$
$$= C_{d_3}A_3((C_{p_3} - C_{p_4})v_z^{\,2})^{\frac{1}{2}}$$

or alternatively

$$(C_{p_1} - C_{p_2})v_z^{\,2} = \frac{Q^2}{C_{d_1}^{\,2}A_1^{\,2}}$$

$$(C_{p_2} - C_{p_3})v_z^{\,2} = \frac{Q^2}{C_{d_2}^{\,2}A_2^{\,2}}$$

$$(C_{p_3} - C_{p_4})v_z^{\,2} = \frac{Q^2}{C_{d_3}^{\,2}A_3^{\,2}}$$

Summing these equations gives

$$(C_{p_1} - C_{p_4})v_z^{\,2} = Q^2 \left(\frac{1}{C_{d_1}^{\,2} A_1^{\,2}} + \frac{1}{C_{d_2}^{\,2} A_2^{\,2}} + \frac{1}{C_{d_3}^{\,2} A_3^{\,2}} \right)$$

or

$$Q = \left(\frac{(C_{p_1} - C_{p_4})v_z^{\,2}}{\dfrac{1}{C_{d_1}^{\,2} A_1^{\,2}} + \dfrac{1}{C_{d_2}^{\,2} A_2^{\,2}} + \dfrac{1}{C_{d_3}^{\,2} A_3^{\,2}}} \right)^{\frac{1}{2}}$$

or generally for more than three openings

$$Q = \left(\frac{(C_{p_1} - C_{p_{n+1}})v_z^{\,2}}{\dfrac{1}{C_{d_1}^{\,2} A_1^{\,2}} + \dfrac{1}{C_{d_2}^{\,2} A_2^{\,2}} + \dfrac{1}{C_{d_3}^{\,2} A_3^{\,2}} + \ldots + \dfrac{1}{C_{d_n}^{\,2} A_n^{\,2}}} \right)^{\frac{1}{2}} \qquad (6.8)$$

This general equation can be simplified in the case where orifice areas are equal or all discharge coefficients are similar.

Application of this and similar equations depends on the availability of suitable wind pressure distribution data as well as appropriate discharge coefficients for a wide range of building openings. The reference wind speed v_z has to be determined for each prevailing wind direction from detailed information on local wind characteristics.

Van Straaten [6.15] has shown that pressure distributions over the surface of solid models do not change significantly when small openings are introduced, except for locations close to the edge of such openings. This allows average pressures on solid models at locations for openings to be used in discharge equations without significant error, provided openings in a wall do not exceed 20% of that wall area. As there are many sources of pressure distributions on solid building forms, data from these sources can offset the need for wind tunnel studies on similar models when wall openings are small.

6.5.3. *Sources of Pressure Coefficient Data for Estimating Natural Ventilation*

There are two major sources of pressure distribution data for building forms, those of Chien, and Jensen and Franck. Models used by Chien [6.16] in pressure distribution studies were simple block forms with none of the architectural features common to many buildings. Studies by Jensen and Franck [6.17] included some models with projecting eaves. Less detailed average pressure coefficients for a range of building forms are provided in wind loading codes [6.18, 6.19]. Significant eaves projections are a common sun control feature on low buildings in warm humid climates. Proportions of such architectural features become very significant on smaller buildings and a common practice of raising the building off the ground can cause appreciable changes in pressure distributions.

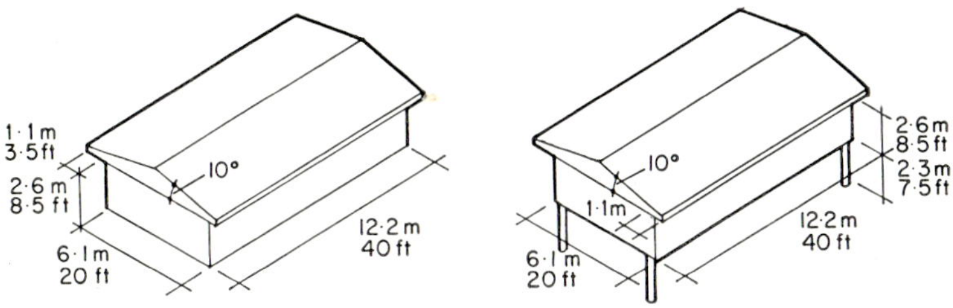

Fig. 6.7. Solid model types used in mid-wall height pressure distribution studies. (Left) low set house; (right) high set house.

Pressure distributions at mid height of walls on two model building forms (Fig. 6.7) [6.20] representing recent housing built in Australia's hot humid tropics are provided in graphical form in Figs. 6.8 and 6.9 for a number of wind incidences to a long wall.

Most pressure distributions on buildings are determined from wind tunnel studies of isolated models. Where a particular building has other buildings or obstructions nearby accurate pressure distributions can only be obtained by modelling such obstructions during the wind tunnel study. In the case of groups of buildings some studies were performed by Bailey and Vincent [6.21].

6.5.4. *Sources of Local Wind Speed Data for Estimating Natural Ventilation*

Data on local wind characteristics can be determined by measurements at the building site which should be of a suitable duration to include appropriate seasonal changes. This is not normally done as, in order to arrive at reliable average values, measurements have to

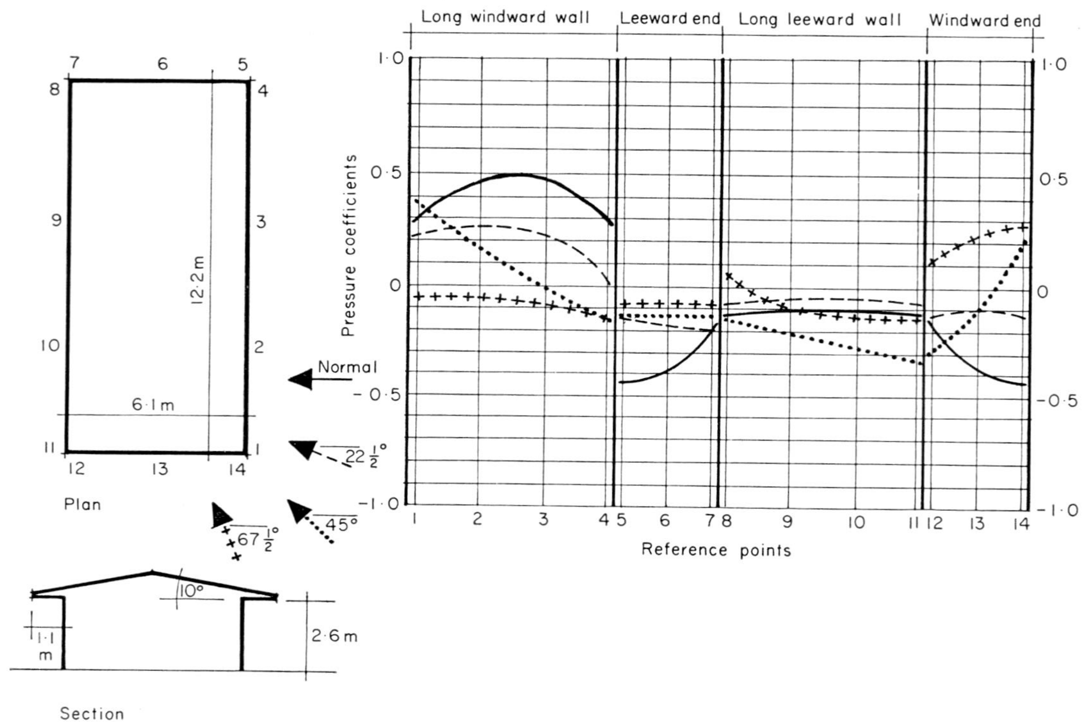

Fig. 6.8. Pressure distribution at midheight of walls on a low set house with eaves. Note. Pressure coefficients above are based on a reference dynamic pressure 10 m above ground. Mean wind speed profile gradient height 400 m and exponent 0·28.

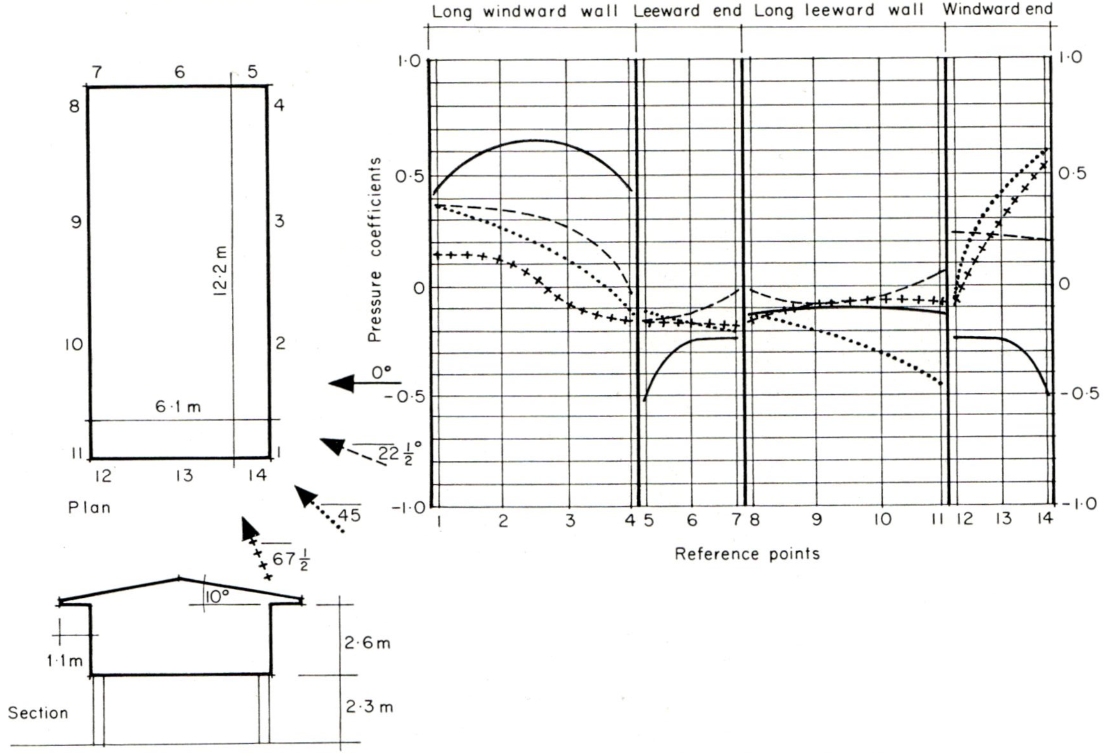

Fig. 6.9. Pressure distribution at midheight of walls on a high set house with eaves. Note. Pressure coefficients above are based on a reference dynamic pressure 10 m above ground. Mean wind speed profile gradient height 400 m and exponent 0·28.

span a number of years. If the building site is close to a meteorological station anemometer and both are surrounded by a similar region of relatively flat terrain, mean velocity records from the anemometer could be expected to indicate those likely at a corresponding height at the site. Small scale (1:2000) topographic models (Fig. 6.10) with

Fig. 6.10. A topographic model scale 1:2000 in University of Western Ontario boundary layer wind tunnel.

appropriately modelled surface roughness have been used to study topographic effects on large scale cyclonic wind. Similar studies could be used to determine approximate relationships between mean wind speeds and directions at a building site and those recorded at the nearest anemometer.

Wind data from Australian meteorological station anemometers is provided in a number of forms. Wind roses (Fig. 6.11) indicate monthly percentage occurrence of calms and wind directions at 9 a.m. and 3 p.m. observed over periods of approximately five years [6.22].

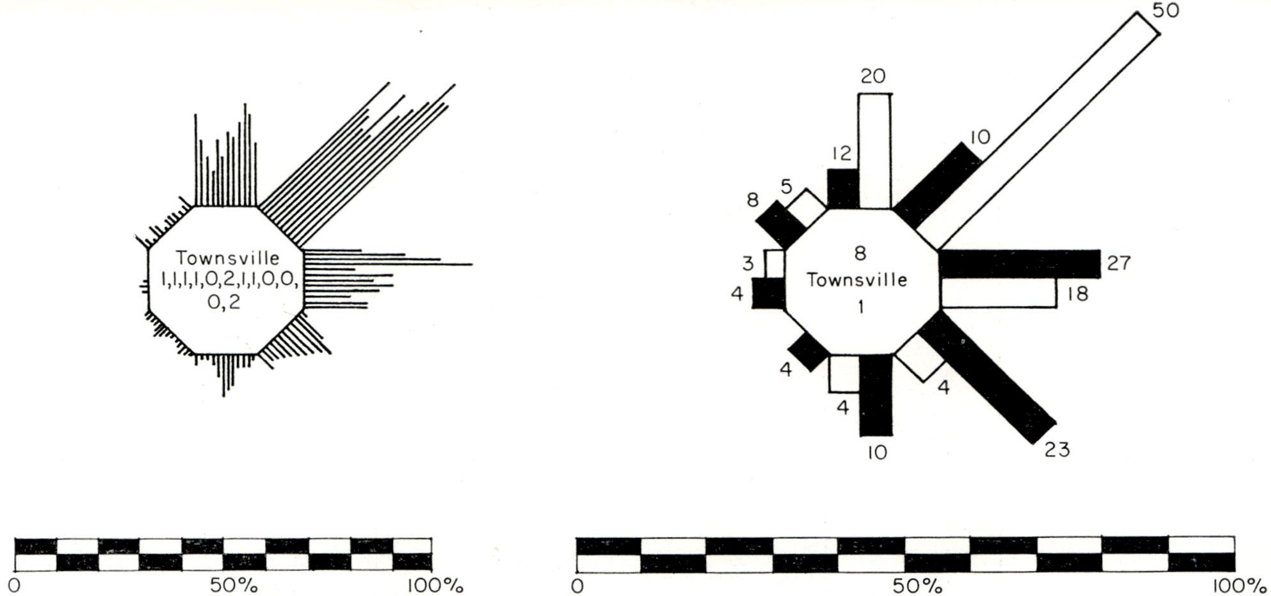

Fig. 6.11. Two types of wind roses indicating wind frequency and direction. (Left) Wind rose for 12 months at 3 p.m. The sides of the octagons face towards the cardinal and semi-cardinal points. Projecting from each side are twelve columns representing the twelve months of the year, and the lengths of the columns are proportional to the percentage frequencies of the winds from the given direction in the successive months, working round clockwise from January to December. The scale is shown above, and the outer octagons have been drawn, separated from the inner octagons by a distance representing $12\frac{1}{2}$%. The tops of the columns would all be on these outer octagons if winds from the eight directions were equally frequent and there were no calms. The percentage frequency of calms in the 12 months is shown by figures within the octagons. Small dashes within the inner octagons indicate months when the percentage of winds from that direction is practically zero. (Right) Wind rose for 3 summer months. Wind direction towards centre of rose. Black bars represent morning winds. Open bars represent afternoon winds. Lengths of bars represent relative frequencies of direction. Numerals on bars indicate percentage of winds from the particular direction which exceed 9 m/s. Figures in the centre circle represent the percentage of calms, morning on top, afternoon below.

Separate tables indicate average wind speeds, highest mean wind speed in one day, and highest gust speed for each month together with prevailing wind directions at 9 a.m. and 3 p.m. (Fig. 6.12) [6.23] for major cities.

	Wind (height of anemometer 10 m)				
Month	Average (km/h)	Highest mean speed in one day (km/h)	Highest gust speed (km/h)	Prevailing direction 9 a.m.	Prevailing direction 3 p.m.
No. of years of observations	56	56	56	30[a]	30[a]
January	12·3	31·8	90	SE	NE
February	12·1	35·7	108	SE	NE
March	11·6	32·7	122	S	E
April	10·5	26·9	103	S	E
May	9·9	28·8	87	SW	SE
June	10·2	30·6	95	SW	WSW
July	9·9	35·4	111	SW	WSW
August	10·2	26·9	100	SW	NE
September	10·5	25·9	101	SW	NE
October	11·1	16·8	100	S	NE
November	11·6	24·9	111	SE	NE
December	12·1	31·4	127	SE	NE
Yearly averages	11·0			SW	NE
Yearly extremes		35·7	127		

[a] Standard thirty years normal (1911–1940).

Fig. 6.12. Wind data: Brisbane, Queensland. Lat. 27° 28′ S., Long. 153° 2′ E. Height above M.S.L. 41 m).

Introduction of computer analysis of wind records has extended this information to computer printout in the form of hourly wind speed and direction frequency data. This data (Fig. 6.13) is the most detailed that is readily available in Australia.

Orientation of the long walls of a building should allow for solar heat gain as well as direction of prevailing breezes. Prevailing winds along the north east coast of Australia, during the hot humid summer months, change direction from the south east sector in the morning, to the north east sector in the evening. To gain maximum benefit

Direction	Calms	Wind speed intervals (m/s) 0·5–1·5	2·0–3·0	3·6–5·0	5·7–8·2	8·7–10·8	Total
		Number of occurrences					
Calms	1						1
NNE		1	4	3	5	–	13
NE		2	7	22	19	–	50
ENE		–	1	18	12	–	31
E		–	1	5	6	3	15
ESE		–	–	1	3	2	6
SE		–	1	3	–	–	4
SSE		–	–	–	–	–	0
S		–	–	–	–	–	0
SSW		–	–	–	–	–	0
SW		–	1	–	–	–	1
WSW		–	–	–	–	–	0
W		–	–	–	–	–	0
WNW		–	–	–	–	–	0
NW		–	–	2	–	–	2
NNW		–	2	3	1	–	6
N		1	6	13	4	1	25
Totals	1	4	23	70	50	6	154

Note: Available from the Bureau of Meteorology as computer printout in knots for each hour of the day and each month of the year (1 knot = 0·5144 m/s).

Fig. 6.13. Wind frequency analysis. (Location: Townsville, North Queensland, Australia; Latitude 19° 15′ S; Longitude 146° 46′ E; Month: January, Time: 1500 hours, Local Time).

from wind in each sector, orientation of the long axis of a building should be either north–south or east–west (Fig. 6.14).

As sunshading on north and south faces can be achieved by projecting eaves and end walls, without obstructing airflow, the north–south orientation of the long walls is clearly superior [6.8] for that geographic region.

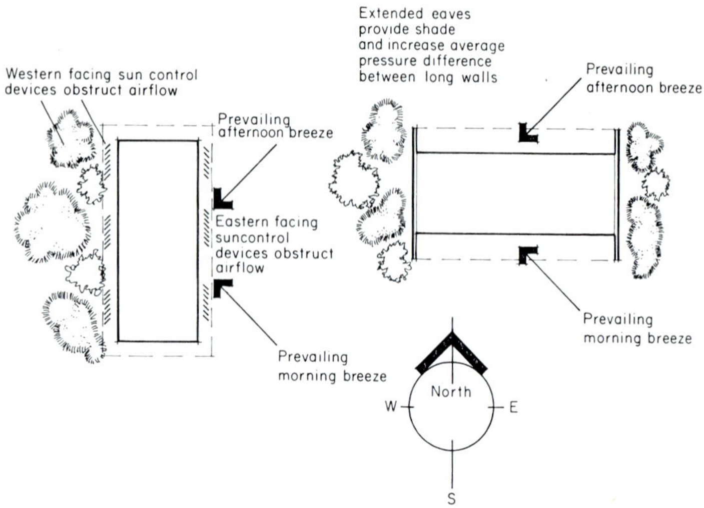

Fig. 6.14. Building orientation considerations for sun control and natural ventilation in Townsville, Australia.

6.5.5. *Sources of Discharge Coefficients for Estimating Natural Ventilation*

Combined discharge coefficients for openings in buildings such as those determined by Van Straaten [6.15] are rather limited considering the variety of typical building openings and their possible sequential combinations. Most discharge coefficients found in texts on fluid mechanics are for single, square-edged, round openings in flat plates, or a range of standard orifice designs used for flow measurement in pipes.

Full scale studies by Dick [6.24] in 1950 established that discharge coefficients of approximately 0·65 were appropriate for very small square edged cracks and openings in houses. A.S.H.R.A.E. Guide [6.14] contains a wide range of both discharge coefficients and dynamic loss coefficients. Discharge coefficients for larger window openings in series were determined by Wannenburg and Van Straaten [6.25] in 1957 for estimating airflow for comfort in school classrooms. These window openings occupied approximately 10% of the wall area. Snyckers [6.26] extended the range of discharge coefficients for a range of rectangular openings in series up to 20% of wall area.

Airflow rates through openings in series estimated using discharge coefficients from Figs. 6.15 and 6.16 were found to be within 15% of measured values in a series of wind tunnel studies on a variety of opening geometries and sequences in which openings did not exceed 10% of the wall area [6.20]. Cosine corrections were made to areas of windward openings in these studies for wind directions not normal to the windward wall.

Description of opening	Typical range of discharge coefficients for normal incidence	Jet characteristics
Small openings in thin walls less than 10% of wall area near the centre of the wall	0·50–0·65	Small inertia due to small mass of air in jet
Openings 10–20% near the centre of a wall with aspect ratio similar to the cross-section of the downwind space	0·65–0·70	Significant inertia due to increased mass of air in jet
Openings 10–20% of a wall with one edge common with the downwind space such as a doorway	0·70–0·80	Wall effect reduces energy losses on one side of jet
Openings similar in size to the cross-section of the downstream space	0·80–0·90	Wall effect around the perimeter of the jet significantly reduces turbulent energy losses

Fig. 6.15. Typical discharge coefficients for single inlet or intermediate openings in buildings

A_0/A_1	C_d
Approaching 0·0	0·63
0·2	0·64
0·4	0·67
0·6	0·71
0·8	0·81
1·0	1·00

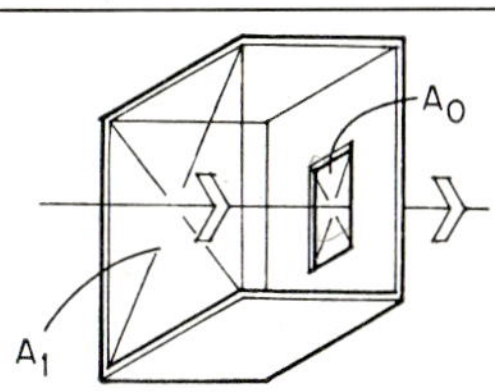

Fig. 6.16. Discharge coefficients for outlet openings.

Investigations by Van Straaten and Snyckers revealed many complex influences of opening geometry on discharge coefficients for rectangular openings in series. Some of the more noticeable influences were:

(i) *Relative shape and sequence of aligned openings.* The effect of downstream openings with smaller dimensions than those of the jet from the upstream opening, is that a significant amount of kinetic energy in the jet is converted to a static pressure increase around the downstream opening. Downstream openings of similar size, or larger than, the jet from a windward opening, result in a much lower energy loss due to less interference with the jet from the windward opening (Figs. 6.17 and 6.18).

(ii) *Alignment of openings with jets from upstream openings* (Figs. 6.19 and 6.20). Alignment of openings with jets ensures minimum energy losses due to forced changes in flow direction.

(iii) *Proximity of surfaces parallel to jets* (Fig. 6.21). Surfaces close and parallel to jets cause the jet to coalesce with the surface. This reduces energy losses due to turbulent mixing at the perimeter of the jet by up to 40% and is referred to as the 'wall effect' [6.14]. Deflection of jets at openings by louvres or high pressure regions on windward faces can accentuate this effect (Fig. 6.22).

(iv) *Downwind spacing of aligned openings* (Fig. 6.23). Close spacing of similar sized, aligned openings can cause a significant reduction in static pressure in the space between the openings [6.15]. This reduction in pressure can result in an increased discharge through the upstream opening. This effect appears to occur when the ratio of the downwind spacing to the hydraulic radius of the opening is less than 80 and is most pronounced when this ratio is less than 30. The hydraulic radius of the duct being the duct area divided by its perimeter. (This situation does not occur frequently in buildings.)

(v) *Depth of reveals or wall thickness in which the opening is located* (Fig. 6.24). Increasing the wall thickness or reveal

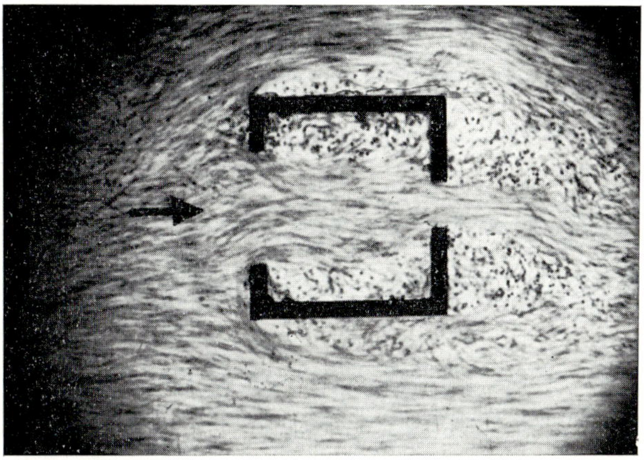

Fig. 6.17. Small downstream openings restrict jet from upstream openings.

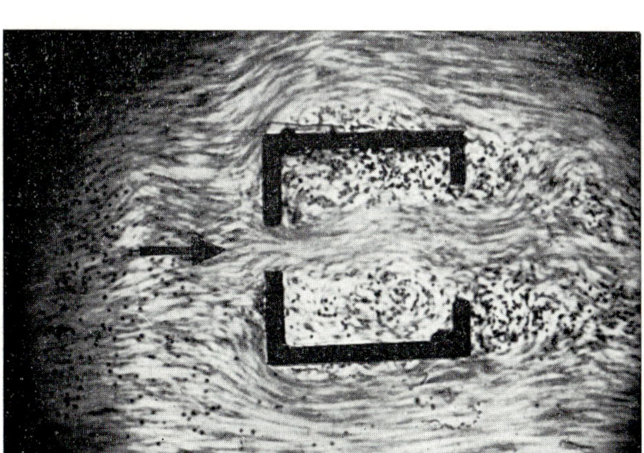

Fig. 6.18. Larger downstream openings minimise interference with jet from upstream openings.

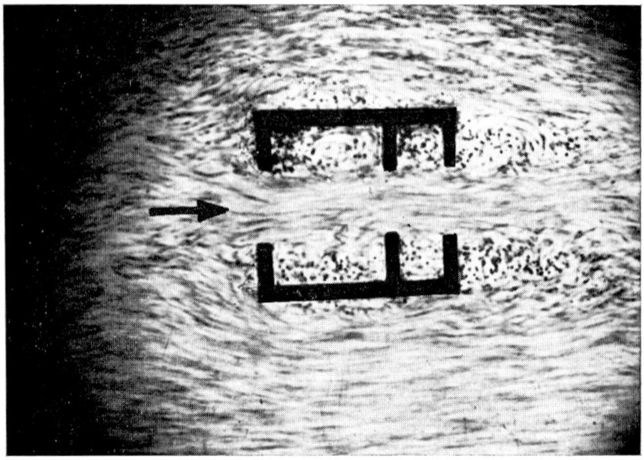

Fig. 6.19. Energy losses minimised by alignment of openings.

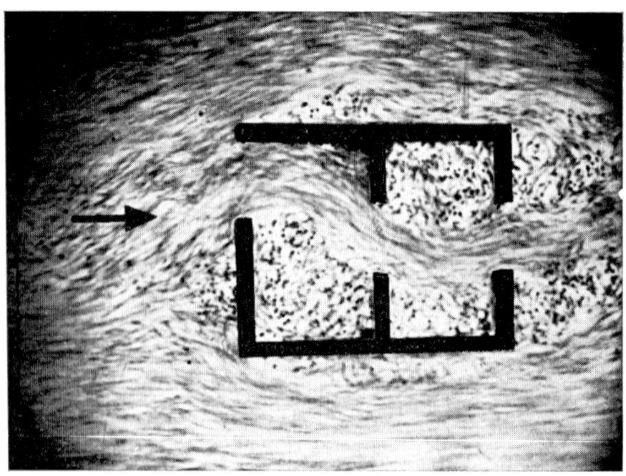

Fig. 6.20. Forced changes in jet direction by unaligned openings.

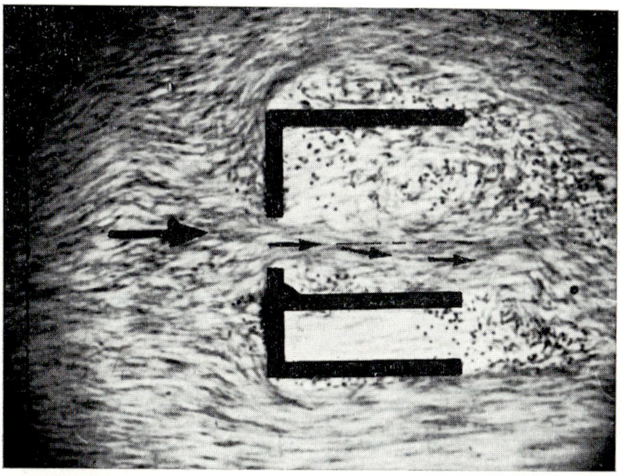

Fig. 6.21. Jet coalescence with parallel wall.

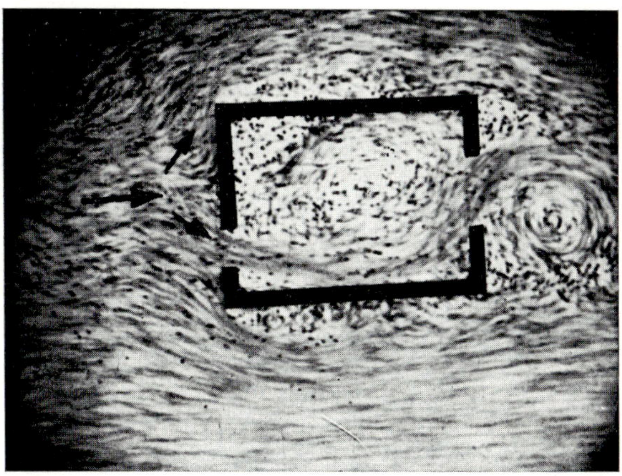

Fig. 6.22. Jet deflection and coalescence due to high pressure region on windward face.

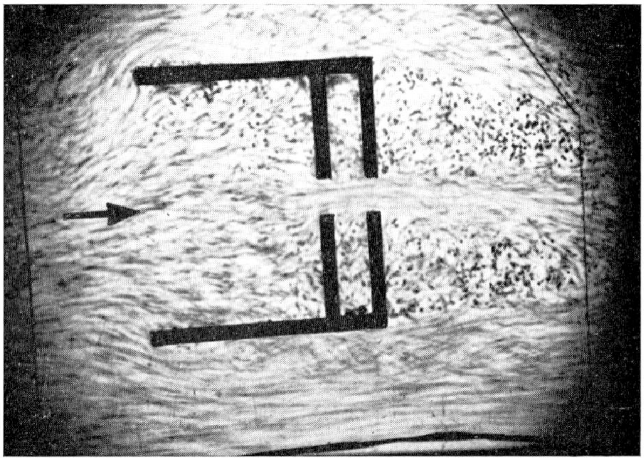

Fig. 6.23. Discharge through windward opening is increased by close proximity to similar aligned downstream opening.

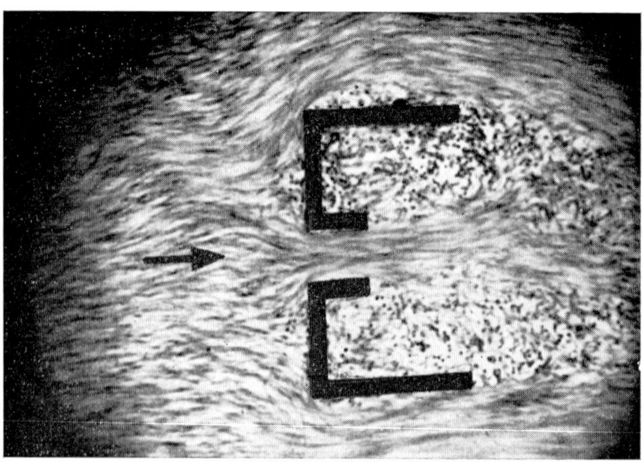

Fig. 6.24. Deep reveals restrict jet expansion reducing turbulent losses in the jet.

depth, has the effect of restricting jet expansion and delaying flow separation from the duct walls. This in turn reduces energy losses associated with turbulent mixing at the perimeter of the expanding jet in downstream spaces.

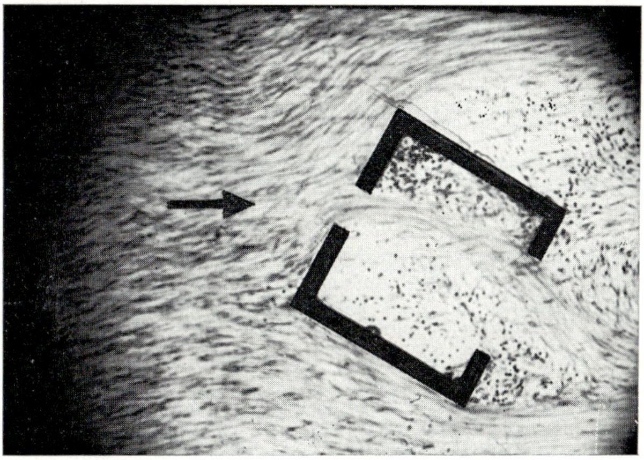

Fig. 6.25. Jets through openings in windward walls inclined to the wind tend to retain the direction of the approaching wind and are more likely to collide with an internal surface before reaching downstream opening.

(vi) *Orientation of windward opening to the wind* (Fig. 6.25). Wind tends to maintain its direction while passing through a clear opening in a windward wall. This has two effects, first the effective area of the opening is reduced by a factor approximating the cosine of the angle of incidence of wind to the wall. Secondly, as rooms are usually rectangular in shape the jet from a windward opening inclined to the wind is more likely to suffer energy losses due to collision with an internal surface of the space before leaving through a downstream opening.

Equation (6.8) obviously does not accommodate the influences described above; however appreciation of these influences will give an indication as to whether the calculated discharge may be larger or smaller than the actual discharge. The effect of the above influences on the combined discharge coefficient for openings in series, can only be determined by wind tunnel measurements and observations.

6.5.6. *Advantages and Disadvantages of Pressure Difference Method for Estimating Natural Ventilation*

Methods for estimating natural ventilation from differences in wind pressure and discharge coefficients have two principal advantages.

(1) Ability to make use of the growing sources of wind pressure distribution data associated with wind load research.
(2) Estimates of natural ventilation can be made without resorting to wind tunnel studies provided suitable pressure distribution and discharge coefficients are known.

These advantages make this method suitable for use in design offices particularly during initial design stages, where such information can assist in evaluating alternative designs. Use of average wind pressure differences at mid wall height from solid model pressure studies at locations where openings are to be provided, give satisfactory estimates of airflow for openings up to 20% of wall area.

Disadvantages of the pressure difference method are:

(1) When openings through a building are in excess of 20% of the wall area, it becomes increasingly difficult to determine the effective pressure difference responsible for airflow through the openings from pressure distributions on solid models.
(2) Estimates of mean internal wind speeds are restricted to locations in the jet close to openings. This can be a significant disadvantage when jets from openings occupy only a small proportion of the occupied space.

To place these disadvantages in perspective, most recently built houses along the hot humid coast of north eastern Australia have wall openings less than 20% of the wall area. The porosity of external walls on residential buildings closely follows the type of windows used. This is due to the adoption of limited range of standard sill and head heights in the home construction industry. Average porosity of the two long walls is approximately:

(1) 20% for typical aluminium sliding windows and doors (Fig. 6.26);
(2) 40% for aluminium sliding windows with louvres below sill and sliding doors (Fig. 6.27);
(3) from 50% to 80% for ceiling height louvres combined with hinged doors (Fig. 6.28).

Fig. 6.26. Residence with aluminium sliding windows and doors. Typical porosity of long walls 20% openings.

Fig. 6.27. Residence with sliding aluminium windows with louvres below the sills and sliding doors. Typical porosity 40% openings.

Internal partition walls rarely have the degree of porosity in external walls. In poorly designed homes where a single doorway is the only internal opening to many rooms average internal wall porosities can be less than 10% and virtually nil if doors are closed. Well planned buildings have louvres in internal partition walls so that porosities approach external wall porosities.

Fig. 6.28. Residence with full height louvre walls. Typical posority 50–80% openings.

The second disadvantage of methods for estimating natural ventilation based on wind pressure differences is more important. As maintenance of indoor comfort in warm humid climates is dependent on air speeds past the occupants it is necessary to be able to estimate air speeds at locations away from openings and outside the main airflow jet. Air speeds in these locations cannot be estimated using wind pressure difference methods. This difficulty can be offset to some extent by arranging furniture in a space to ensure occupants are located in air jets close to openings or by adjusting louvres in the opening to direct the airflow jet into occupied areas.

Example

Determine, using the discharge equation, the extent to which natural ventilation is likely to maintain comfort in the living room of the isolated high set house (Fig. 6.29) in Townsville at 3 p.m. during January.

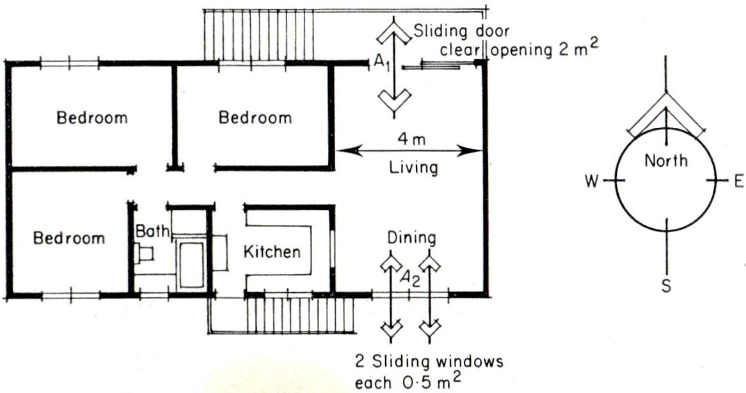

Fig. 6.29. Plan of a high set (elevated) house showing opening to living room.

Climatic data

3 p.m., 86 percentile average maximum dry bulb temperature = 32·9°C. 3 p.m., average relative humidity = 62%.

Assuming adequate insulation is provided in the building envelope and windows and walls are shaded, the minimum airflow needed to restore thermal comfort for the above temperature and humidity from Fig. 6.2 is 0·83 m/s.

	Calms	Wind speed intervals (metres per second)					
Direction		0·5–1·5	2·0–3·0	3·6–5·0	5·7–8·2	8·7–10·8	Total
	0·6%						0·6%
NNE		0·6%	2·6%	1·9%	3·2%	—	8·4%
NE		1·3%	4·5%	14·3%	12·3%	—	32·5%
ENE		—	0·6%	11·7%	7·8%	—	20·1%
E		—	0·6%	3·2%	3·9%	1·9%	9·7%
ESE		—	—	0·6%	1·9%	1·3%	3·9%
SE		—	0·6%	1·9%	—	—	2·6%
SW		—	0·6%	—	—	—	0·6%
NW		—	—	1·3%	—	—	1·3%
NNW		—	1·3%	1·9%	0·6%	—	3·9%
N		0·6%	3·9%	8·4%	2·6%	0·6%	16·2%

Fig. 6.30. Percentage distribution of 10 m winds at 3 p.m. at Townsville during January.

Wind speed frequency and directions for Townsville at 3 p.m. during January are indicated in Fig. 6.13. It would be reasonable to expect this data to be representative of most Townsville suburbs as there is no significant change in terrain over most of the suburban area.

Wind frequency data from Fig. 6.13 is expressed as percentages of total observations in Fig. 6.30.

Building data

Orientation and layout of isolated high set house are as indicated in Fig. 6.29.

Openings in living room walls:

	Dimensions	Clear area of opening
Sliding door north facing wall	2 m × 1 m	$2{\cdot}0\ m^2$
2 sliding windows south facing wall	1 m × 0·5 m each	$1{\cdot}0\ m^2$
Cross-section of living room	3 m × 4 m	$12\ m^2$

Method

As the mid wall height pressure distribution data in Fig. 6.9 is only available for normal, 30° and 45° incidence, the principal wind directions were grouped as follows:

Wind direction	Incidence used to obtain pressure differences
NNE	30° (in lieu of $22\frac{1}{2}$°)
NE and ENE	45°
E	No significant pressure difference
ESE and SE	45°
SW	45°
NW	45°
NNW	30° (in lieu of $22\frac{1}{2}$°)
N	0°

Volumetric discharge is calculated for each wind direction using the lower wind speeds of each wind speed interval commencing from 0·5 m/s. This volumetric discharge is divided by the area of the windward and leeward openings in turn to estimate the average wind

Wind direction	Incidence θ	C_{p1}	C_{p2}	A_1	$A_1 \cos\theta$	A_2	$A_2 \cos\theta$	C_{d_1}	C_{d_2}	Minimum wind speed interval for thermal comfort	% occurrence of thermal comfort
NNE	22½	0·36	0	2	1·8	1	—	0·75	0·63	5·7	3·2
NE	45	0·30	−0·38	2	1·4	1	—	0·75	0·63	3·6	26·6
ENE	67½	0·14	−0·07	2	0·77	1	—	0·75	0·63	5·7	7·8
ESE	67½	−0·07	0·14	2	—	1	0·38	0·64	0·70	14·4	—
SE	45	−0·38	0·30	2	—	1	0·71	0·64	0·70	5·7	—
SW	45	−0·16	0	2	—	1	0·71	0·64	0·70	11·3	—
NW	45	0·0	−0·16	2	1·4	1	—	0·75	0·63	5·7	—
NNW	22½	0·16	−0·07	2	1·8	1	—	0·75	0·63	5·7	0·6
N	0	0·55	−0·12	2	2·0	1	—	0·75	0·63	3·6	11·6
									Total		49·8 %

Fig. 6.31.

speed through the opening. When an average wind speed through both windward and leeward opening is reached that exceeds the minimum necessary to restore thermal comfort in the space, the percentage frequency of occurrence for that and all higher wind speed intervals for a particular wind direction are summed. Repeating this process for each wind direction and totalling all percentages of occurrence indicates the total percentage of time when natural ventilation is likely to be able to restore thermal comfort, in the living room at 3 p.m. during January.

A typical calculation for NE winds is provided below. Results for other wind directions are indicated in Fig. 6.31.

NE winds

Wind incidence used, 45°.

Pressure coefficients at locations near openings from Fig. 6.9:

Windward $+0{\cdot}33$
Leeward $-0{\cdot}38$

Free area of windward opening $= 2\ \text{m}^2$

Correction factor for 45° incidence = cosine 45° = 0·707

Corrected free area $= 2\times 0{\cdot}707$
$= 1{\cdot}4\ \text{m}^2$

Discharge coefficient from Fig. 6.15 for windward opening with two edges common with downwind space = 0·75.

Free area of leeward openings $= 1\ \text{m}^2$.

Discharge coefficient for leeward openings from Fig. 6.16 with an A_0/A_1 ratio of 1/12th or 0·08 = 0·63.

Substituting these values into the discharge equation

$$Q = \left(\frac{(C_{p_1}-C_{p_2})v_{10}^{\,2}}{\dfrac{1}{C_{d_1}{}^2A_1{}^2}+\dfrac{1}{C_{d_2}{}^2A_2{}^2}}\right)^{\frac{1}{2}}$$

Minimum wind speed interval to restore comfort is 3·6–5·0 m/s. As can be seen by substituting 3·6 m/s for V_{10} the external wind speed 10 m above ground level.

$$Q = \left(\frac{(0{\cdot}30-(-0{\cdot}38))3{\cdot}6^2}{\dfrac{1}{0{\cdot}75^2\times 1{\cdot}4^2}+\dfrac{1}{0{\cdot}63^2\times 1^2}}\right)^{\frac{1}{2}}$$

$$= 1{\cdot}60\ \text{m}^3/\text{s}$$

$$\text{Average wind speed per windward opening} = \frac{Q}{A_1\cos\theta}$$

$$= \frac{1{\cdot}60}{1{\cdot}4}$$

$$= 1{\cdot}14\ \text{m/s} > 0{\cdot}83\ \text{m/s}$$

Wind speed through windward opening is sufficient for comfort.

$$\text{Average wind speed through leeward openings} = \frac{Q}{A_2}$$

$$= \frac{1 \cdot 60}{1}$$

$$= 1 \cdot 60 \text{ m/s} > 0 \cdot 83 \text{ m/s}$$

Wind speed through leeward openings is sufficient for comfort.

Percentage occurrence for this 3·6–5·0 m/s wind speed interval and higher intervals for north east wind direction from Fig. 6.30 is: NE 14·3+12·3 = 26·6% of time.

When all wind directions are considered, comfort is likely to be achieved near both windward and leeward openings for dry bulb temperature of 33°C and relative humidity of 62% approximately, for 50% of the time at 3 p.m. during the month of January.

6.6. Velocity Coefficient Method of Estimating Natural Ventilation

Another method of assessing natural airflow inside buildings adopted by most texts on the design of buildings for hot humid climates involves the use of velocity coefficients. These coefficients C_v are the ratio of the mean air speed at a point of interest $\bar{v}_1$ to the mean wind speed $\bar{v}_z$ at a specified reference height upstream from the building in the undisturbed flow so that:

$$C_{\bar{v}_1} = \frac{\bar{v}_1}{\bar{v}_z}$$

By adopting a common reference wind speed in similarly modelled winds, air speeds in different building models can be directly compared. Many velocity coefficient studies in the past made use of reference heights such as sill or eaves height [6.13]. When using low reference heights, it is important to ensure they are not influenced by local obstructions in the flow such as other building models or landscaping elements.

With low buildings such as single or two storey houses, a more

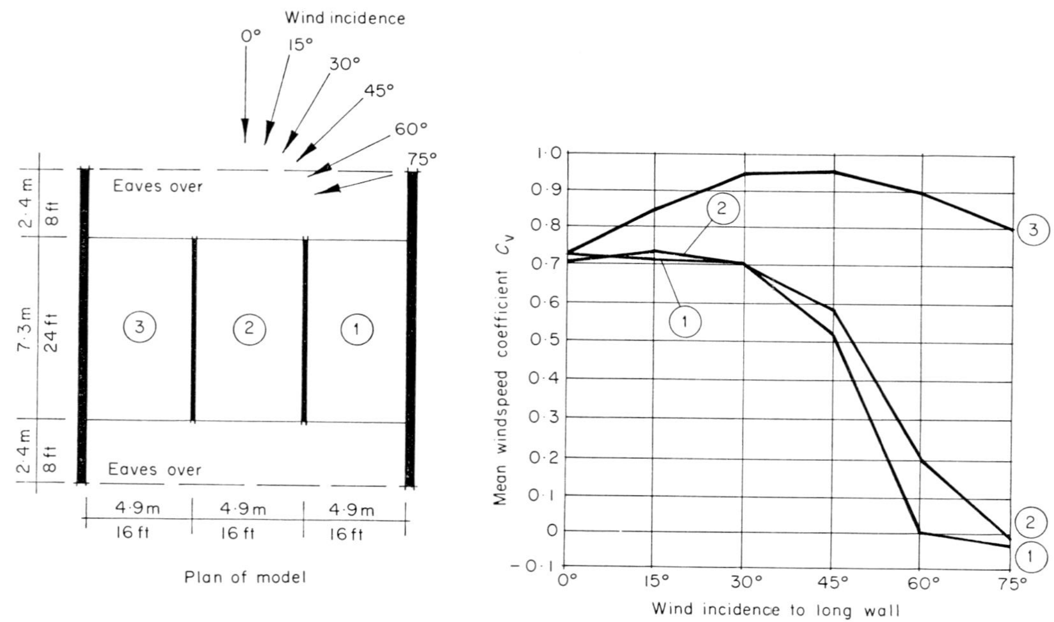

Fig. 6.32. Mean wind speed coefficients at varying incidence through a low set house with extended eaves and end walls and two internal partitions. Note. Reference wind speed for wind speed coefficients was the mean wind speed at a height of 10 m in a mean wind speed profile with a gradient height of 400 m and an exponent of $\alpha = 0{\cdot}28$.

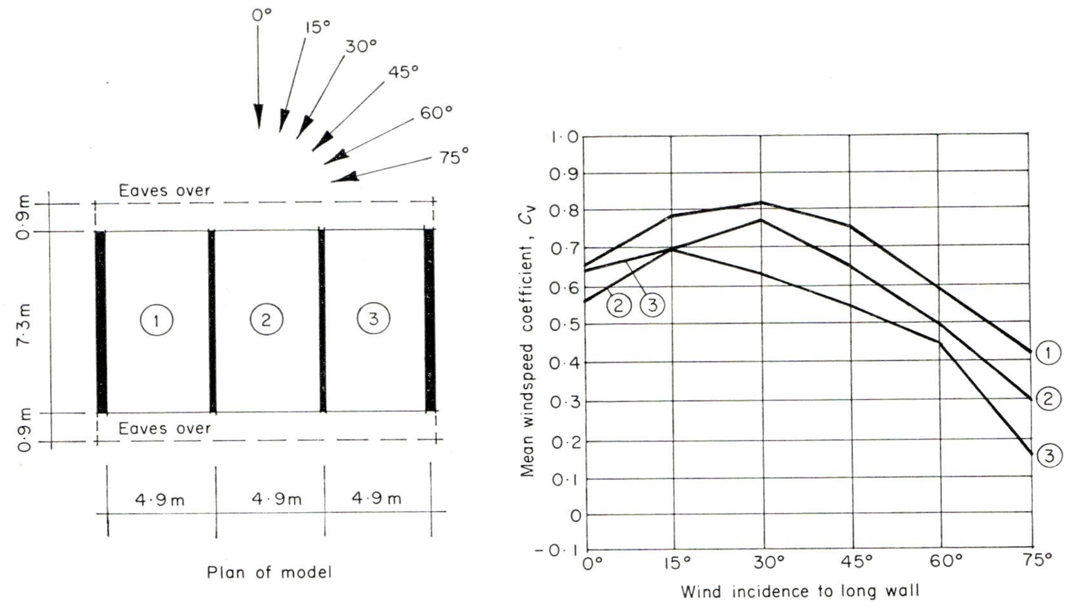

Fig. 6.33. Mean wind speed coefficients at varying incidence through a low set house with small eaves and two internal partitions. Note. Reference wind speed for wind speed coefficients was the mean wind speed at a height of 10 m in a mean wind speed profile with a gradient height of 400 m and an exponent $\alpha = 0{\cdot}28$.

useful reference is that at a height of 10 m above the ground. Mean wind speeds at this height can be more easily related to meteorological station wind records for the area provided appropriate terrain and velocity profiles have been modelled. When the mean 10 m wind speeds $\bar{v}_{10}$ and direction near the building site are available for a particular month and time of day of interest, an air speed $\bar{v}_n$ in or around the full scale building can be estimated. Using the appropriate velocity coefficient C_{v_n} from a wind tunnel study with the model orientated at the corresponding wind direction:

$$\text{Full scale } \bar{v}_n = C_{v_n}\bar{v}_{10}$$

Velocity coefficients for mean jet velocities related to 10 m reference wind speeds for highly porous buildings subdivided by solid partitions determined by Aynsley [6.20] are indicated in Figs. 6.32 and 6.33. Airflow patterns indicating locations and relative dimensions of jets for normal, 30° and 45° incidence, are indicated in Figs. 6.34 to 6.39. The influence of extended end walls on flow pattern is clearly visible. Mean air speeds through these models at normal incidence were the same as external free wind speeds at corresponding

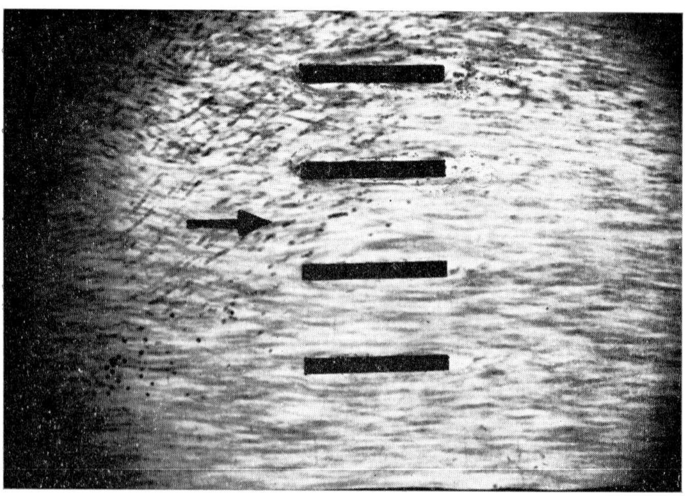

Fig. 6.34. Flow pattern through a highly porous building with three equal cross ventilated compartments. Normal incidence.

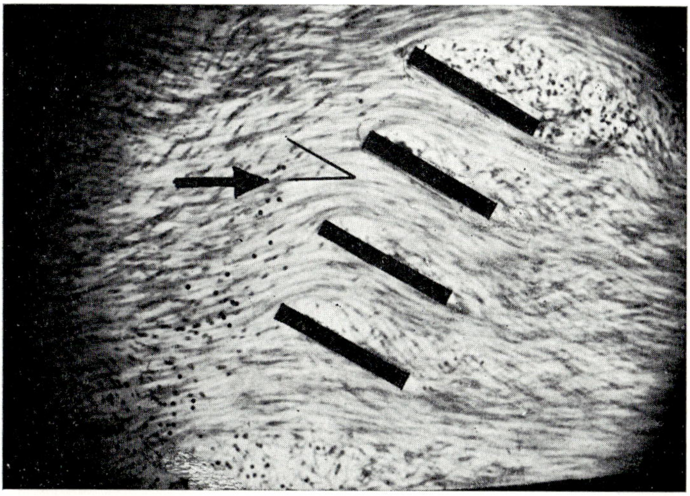

Fig. 6.35. Flow pattern through a highly porous building with three cross ventilated compartments. 30° incidence.

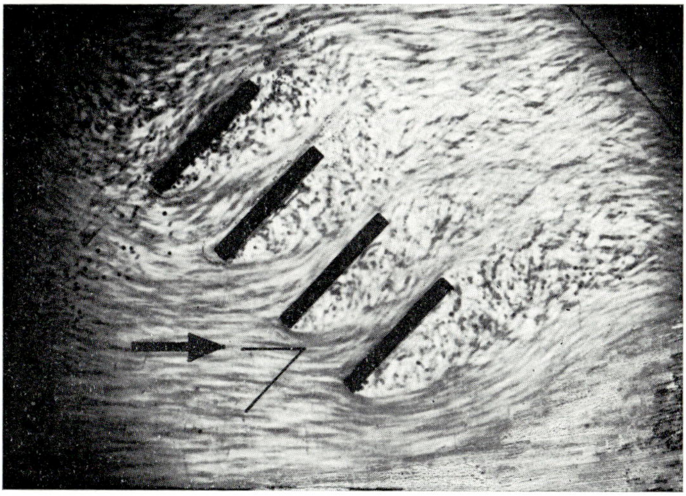

Fig. 6.36. Flow pattern through a highly porous building with three equal cross ventilated compartments. 45° incidence.

Fig. 6.37. Flow pattern through a highly porous, cross ventilated building with three compartments and extended end walls. Normal incidence.

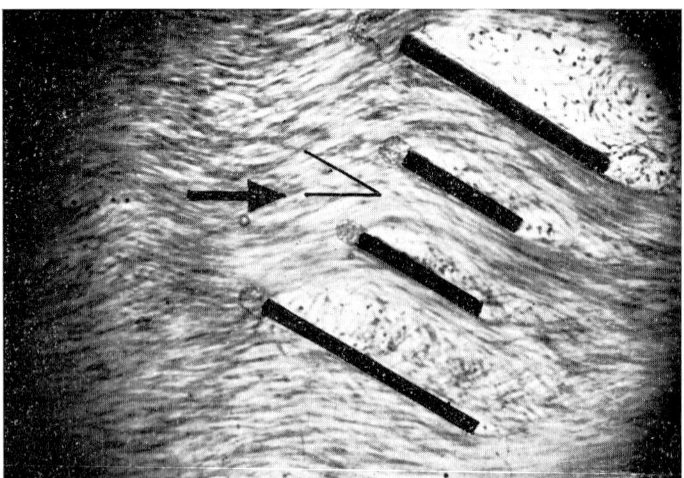

Fig. 6.38. Flow pattern through a highly porous, cross ventilated building with three compartments and extended end walls. 30° incidence.

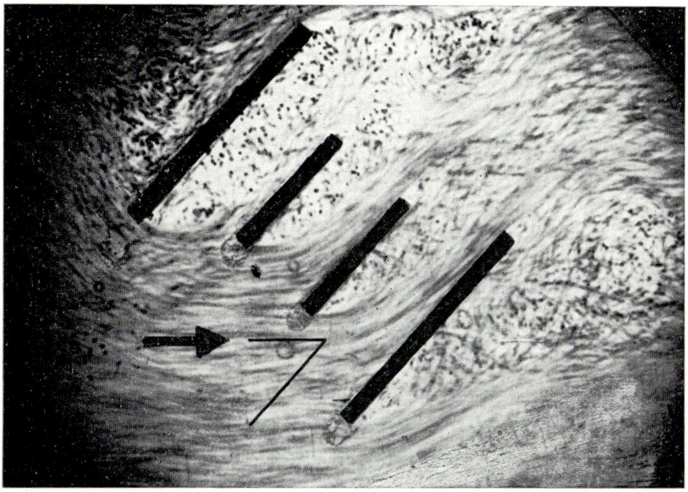

Fig. 6.39. Flow pattern through a highly porous, cross ventilated building with three compartments and extended end walls. 45° incidence.

heights. Velocity coefficients greater than those for normal incidence on these models, indicate increases in air speed in jets above free external wind speeds at corresponding heights.

The simplicity of the velocity coefficient method for estimating air velocities in buildings is appealing; however, the time and expense involved in wind tunnel studies to determine velocity coefficients is restrictive. Most studies have been made in research programmes at government building research stations and at universities.

6.6.1. Sources of Velocity Coefficients and Typical Applications

Staff at the Texas Engineering Experiment Station performed numerous ventilation studies of houses and school classrooms [6.10, 6.27 and 6.28] which were published in 1951.

Natural ventilation of industrial buildings was studied by Weston [6.12 and 6.29] in the wind tunnel at the Commonwealth Experimental Building Station in Australia during 1954. Further wind tunnel studies by Van Straaten [6.9] of school classrooms provided additional velocity coefficient data. More recent sources of velocity coefficients for housing come from studies by Givoni [6.2 and 6.13] and Chand [6.30].

Experiments by Smith [6.31] on models and at full scale at the Texas Engineering Experiment Station in 1951 showed wind tunnel models of buildings could be used to study natural ventilation using velocity coefficients provided they were of a sufficient scale to allow accurate detailed modelling of opening geometry. Another important modelling criterion is wind tunnel blockage. This can increase pressure differences between windward and leeward faces of models. Model blockage should be kept to less than 3% of wind tunnel cross-section to avoid the need for large corrections [6.32].

Results of studies by Givoni [6.13] showed, for a given size of windward opening, higher airflow rates are achieved with larger openings in the leeward side of the space. Arrangements with larger openings in windward walls reduced airflow rates to less than occurred when windward and leeward openings were the same size. Implications of these effects are that where local changes in prevailing winds of similar velocity result in a reversal of airflow direction through a building, openings should be equal in size in opposite walls.

The velocity coefficient approach is ideally suited to study airflow through and around buildings of complex or unusual shape. A wind pressure difference approach of such a building would require a detailed wind tunnel study of wind pressure distributions over the model surfaces. With a wind tunnel study required, it is much simpler to measure velocity coefficients directly, that is unless some other use can be made of the pressure distribution data such as in wind loading considerations.

A typical natural ventilation study in this category, is indicated in Fig. 6.40, in the Department of Architectural Science's wind tunnel at the University of Sydney. The building was a large single storey building designed for a location in the central Pacific. Most external walls consisted of floor to ceiling sliding screens and extensive use was made of quadrant-shaped ventilating hoods fixed at various locations on the roof. The design also incorporated a number of courtyard areas within the building to which wall openings were provided. Velocities inside this model, at a height equivalent to 1 m above floor and at a 10 m reference height were measured with a thermocouple type anemometer. Output from the anemometer was recorded using an x–y plotter on paper calibrated to the anemometer response (Fig. 6.41).

Fig. 6.40. Wind tunnel model of an unusual building used to determine velocity coefficients for estimating natural ventilation.

Model scales for this type of velocity coefficient study should be such that accurate modelling of openings can be achieved. This is particularly important where louvres, pivoted sashes or sun shading devices are involved.

Example

Determine the likelihood of maintaining comfort by natural ventilation in the living area of a low set residence with extended eaves and end walls (Fig. 6.42) in Townsville at 3 p.m. during January.

Climatic data

3 p.m. 86 percentile avg. max. dry bulb temperature	= 32·9°C
3 p.m. average relative humidity	= 62%
Airflow needed to restore comfort from Fig. 6.2	= 0·83 m/s

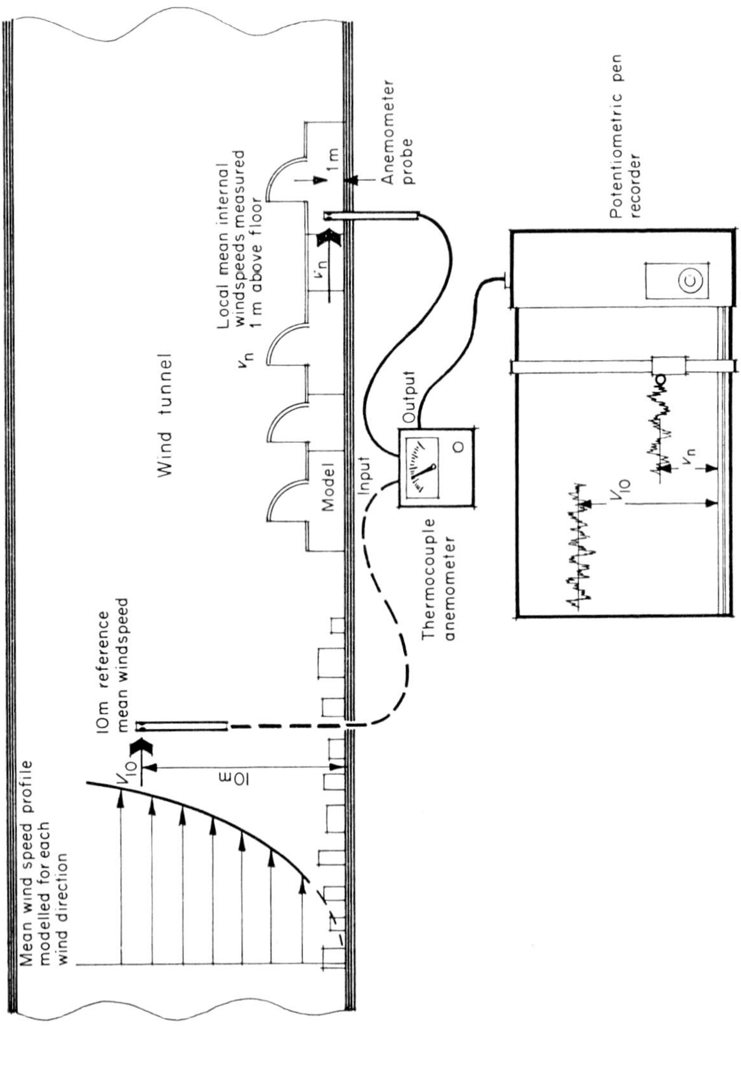

Fig. 6.41. Wind tunnel technique for determining mean wind speed coefficients. Mean wind speed coefficient $C_{v_n} = v_n/V_{10}$. Local mean wind speeds v_n are estimated from 10 m wind records $v_n = C_{v_n} \times V_{10}$.

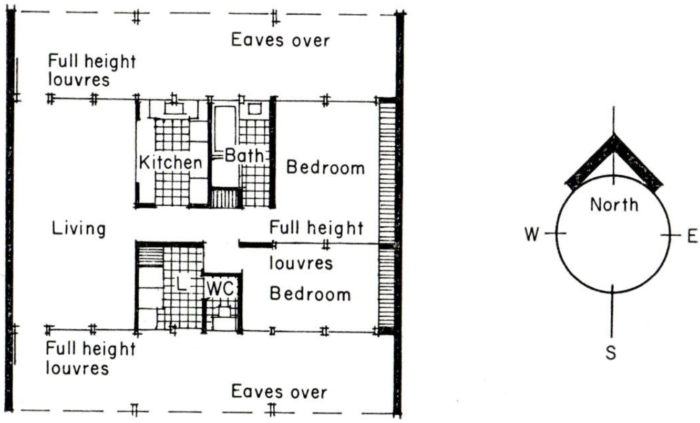

Fig. 6.42.

Building data

Velocity coefficients taken from Fig. 6.32 for jets indicated in Figs. 6.37–6.39.

Method

Using the wind frequency data expressed in percentage of occurrence, from Fig. 6.30, the lower velocity limit of each wind speed increment is multiplied by the velocity coefficient for the living area at wind incidence associated with each wind direction. As in the previous example based on pressure difference, these estimates of internal wind speeds are calculated beginning with the lowest wind speeds until a wind speed increment is encountered that results in internal wind speeds that equal or exceed that minimum required to maintain thermal comfort. Percentage of occurrence for this and all higher increments of wind speed for that direction are then summed. This process is repeated (Fig. 6.43) for each wind direction until a total percentage of time during which comfort is likely to be maintained is obtained.

Actual % of time comfort can be maintained will be higher than 87·0% as wind speeds through living area with a 2 m/s external reference wind speed all exceed the minimum wind speed for thermal comfort of 0·83 m/s by a significant amount. However, the size of wind speed increments provided in wind data prevent a more accurate appraisal.

Direction	Velocity coefficient C_v from Fig. 6.32	Wind speed increments 0·5–1·5: $C_v \times 0{\cdot}5$ (m/s)	Wind speed increments 2·0–3·0: $C_v \times 2{\cdot}0$ (m/s)	Frequency of winds able to restore indoor comfort (%)
NNE	0·9	0·45 (<0·83)	1·8 (>0·83)	7·8
NE	0·95	0·48 (<0·83)	1·9 (>0·83)	31·2
ENE	0·85	0·43 (<0·83)	1·7 (>0·83)	20·1
E	negligible	—	—	—
ESE	0·85	0·43 (<0·83)	1·7 (>0·83)	3·9
SE	0·95	0·48 (<0·83)	1·9 (>0·83)	2·6
SW	0·52	0·26 (<0·83)	1·04 (>0·83)	0·6
NW	0·52	0·26 (<0·83)	1·04 (>0·83)	1·3
NNW	0·71	0·36 (<0·83)	1·42 (>0·83)	3·9
N	0·7	0·35 (<0·83)	1·4 (>0·83)	15·6
			Total	87·0%

Note. Minimum airflow necessary to restore indoor thermal comfort with dry bulb temperature 33°C and relative humidity 62% is 0·83 m/s from Fig. 6.2.

Fig. 6.43. Schedule of percentage frequency of 10 m winds able to restore thermal comfort.

It can be seen by comparing this result of 87% with the previous example of 67% for the high set house described in Section 6.5.6 that highly porous buildings of this type offer more thermal comfort, than less porous conventional designs even when raised some distance above the ground.

6.6.2. *Advantages and Disadvantages of Velocity Coefficient Method*

The principal advantages of the velocity coefficient approach to estimating natural ventilation are:

(1) Simplicity of the method, only reference and internal mean velocities are measured.
(2) The ability to obtain velocity coefficients for any point in the model, allows estimates of air velocities and corresponding thermal comfort to be determined in main airstreams as well as in sheltered eddy zones.

Disadvantages of the method are:

(1) Need for a wind tunnel study of each model of interest.
(2) Need for extremely accurate modelling where complex openings are involved.
(3) Difficulty in estimating full scale velocities when height for external reference velocities does not correspond with height for local wind speed records.

6.7. Design Considerations for Airflow Patterns Inside Buildings

Internal airflow patterns are of vital importance where air movement is intended to improve the thermal comfort of occupants. Airflow within the building is governed largely by the external wind direction, geometry at the entry and inertia forces in the jet which resist changes in direction of the jet. The influence of outlet openings on internal airflow patterns is small except in regions very close to the outlet.

In the case of openings in a windward wall which are small compared to the wall size, the velocities of the jet will be much higher than velocities outside the jet in the eddy flow. Many studies have been made of internal flow patterns, but the general flow pattern can usually be predicted by considering the likely jet behaviour in a space [6.33]. One measure of internal flow often used is the average air velocity at a horizontal reference plane in a space. Another is confined to velocities at selected points along jets. Airflow within the first 2 m above floor level is of greatest interest as this represents the occupied zone. The design of a room and its openings will depend on whether local high velocities are required to amplify external wind speed over small parts of the room or whether the highest possible average wind speed is required across the whole room. The latter condition provides much greater flexibility for the location of various activities in the space.

Airflow above head height is of little use for cooling occupants; however directional control of high level airflow is easily achieved by the use of louvres or other deflecting devices. Louvres in the upper portions of external walls allow airflow through them to be effectively utilised (Fig. 6.44). A practical problem often encountered when

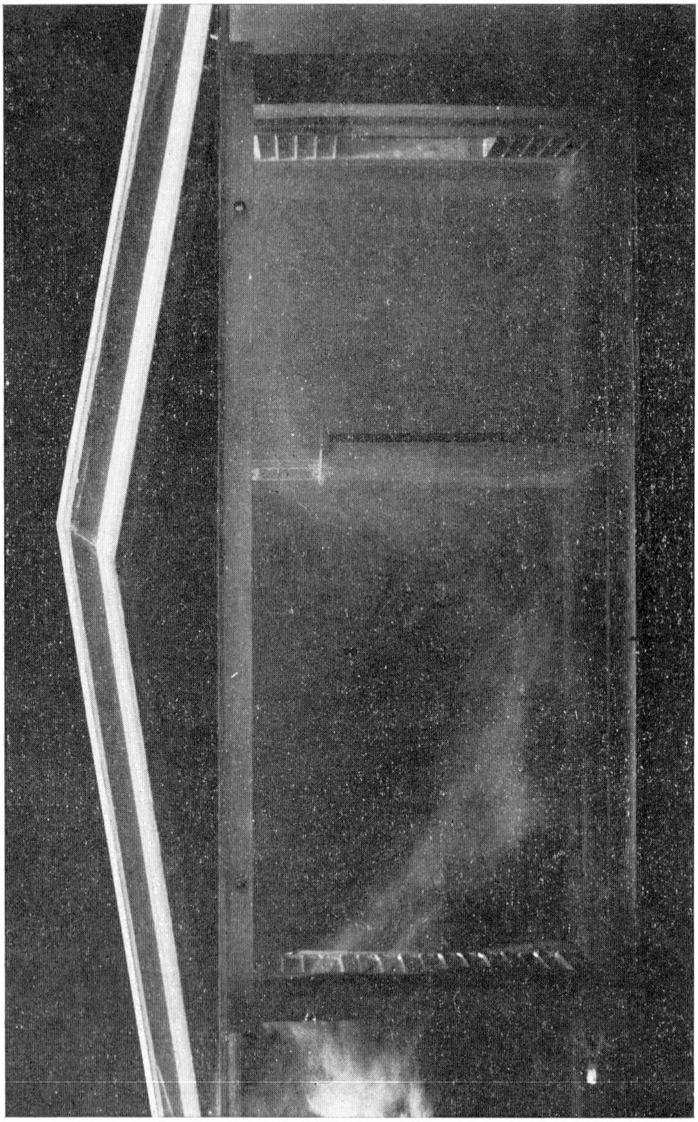

Fig. 6.44. Louvres can be an effective means of directing air jets into occupied areas.

using louvres to deflect airflow downwards in occupied areas is that most commercially available louvres cannot rotate much more than 90° and must be mounted in reverse. Mounting louvres in this way can allow rain penetration but since most tropical housing has large eaves projections this consideration loses importance. Methods in which windows are opened varies considerably and can have significant effects on airflow through the opening. Resistance to

Screen material	Velocity through clear opening (m/s)	Velocity reduction with screen (%)
Bronze wire screen	2	26
5·5 wires/cm	4	17
porosity 80%	6	11
Plastic coated fibreglass	2	41
7 threads/cm	4	34
porosity 66%	6	19

Fig. 6.45.

airflow and the direction of the jet into a room vary with window type, particularly with awning, pivot, or casement designs. Characteristics of airflow patterns through a range of window types were studied by Holleman [6.28] and Van Straaten [6.9].

Obstructions to airflow around the perimeter of a building can have significant effects on airflow inside the building. Some of these effects due to landscaping have been investigated by White [6.34] and Van Straaten [6.15]. More general data on shelter effects of trees is provided by Geiger [6.35].

Use of insect screening can significantly reduce airflow through openings. Resistance to airflow through insect screens varies roughly as the cosine of its angle of incidence to the wind and also with wind speed. Velocity reductions through openings due to insect screens for a range of wind speeds normal to a screen are indicated in Fig. 6.45.

Many residents in Northern Australia value the improved airflow without screening more highly than the insect protection afforded by screens. Mosquitoes present the main problem and burning of insect

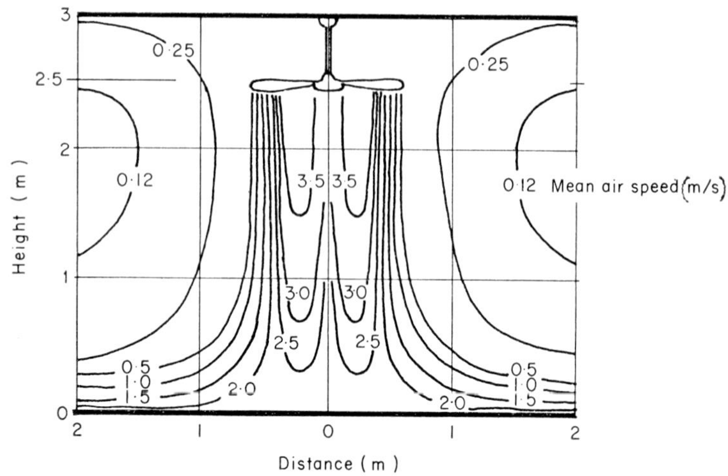

Fig. 6.46. Mean air speed (m/s) distribution for ceiling fan 0·91 m diameter.

repellent coils is often used as a substitute for screening. Another alternative to an insect screen is the use of large diameter slow revolving ceiling fans (Fig. 6.46). The airflow from these fans provides comfort in the absence of wind and flying insects find it difficult to fly in the highly turbulent flow from these fans.

6.8. Flow Visualisation Techniques

Any study of internal airflow velocities can be made more meaningful when combined with flow visualisation studies. Care must be taken, however, to ensure that the techniques used indicate realistic flow patterns. The most reliable methods are those suitable for use on wind tunnel models in carefully modelled boundary layers.

Smoke streams are rapidly diffused in turbulent boundary layers and where sufficient smoke density is achieved it often obscures flow patterns in other areas of interest. Internal flow can be studied through models constructed of clear plastic (Fig. 6.44) by allowing smoke to pass through the model with the flow. Alternatively the interior of the model can be filled with smoke before the airflow is commenced. This later technique shows more clearly the low velocity eddy zones in which the smoke is retained as opposed to faster moving

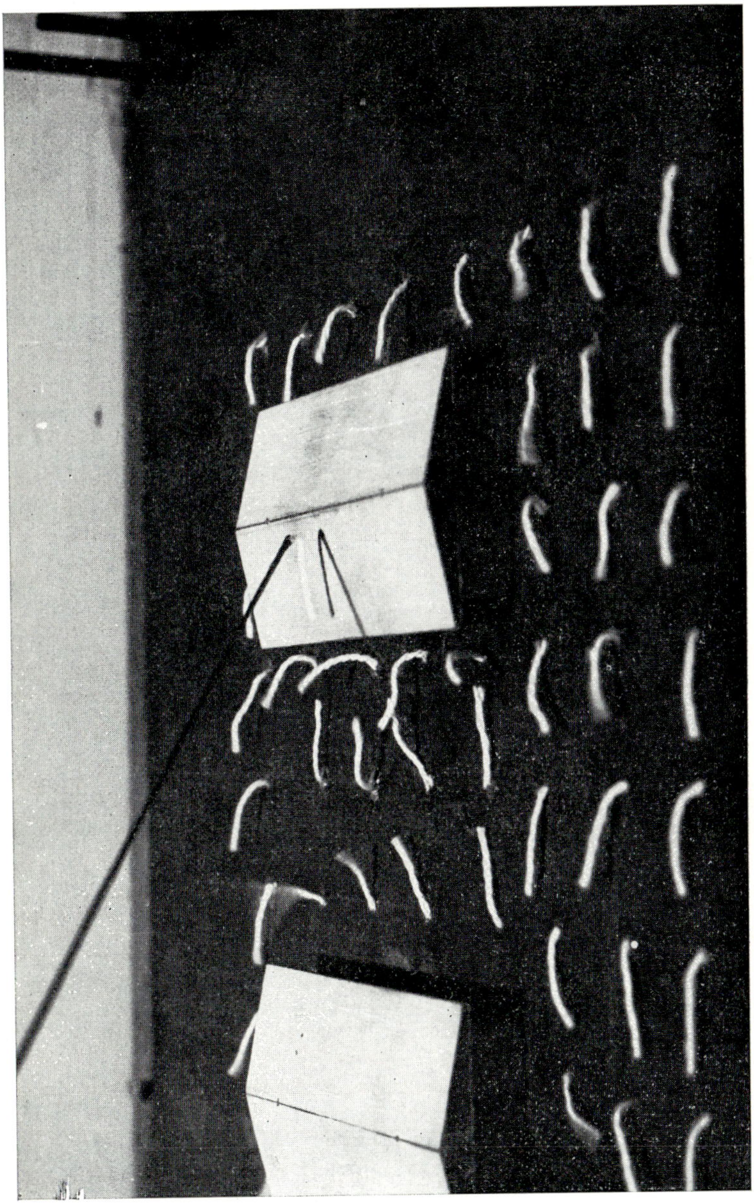

Fig. 6.47. Flow visualisation around tropical housing models using wool tufts.

air jets from which smoke is quickly removed. Use of smoke or dense white fumes produced by reaction of titanium tetrachloride with water vapour in the air, can cause discomfort in enclosed spaces and in some cases constitutes a hazard to health or increases fire risk.

Use of tufts of threads with one end attached to model surfaces or fine wires fixed between the surfaces of a space present a pollution free method for flow visualisation. This technique is also useful for flow visualisation of airflows around buildings to determine changes in flow pattern responsible for pressure changes (Figs. 6.47 and 6.48).

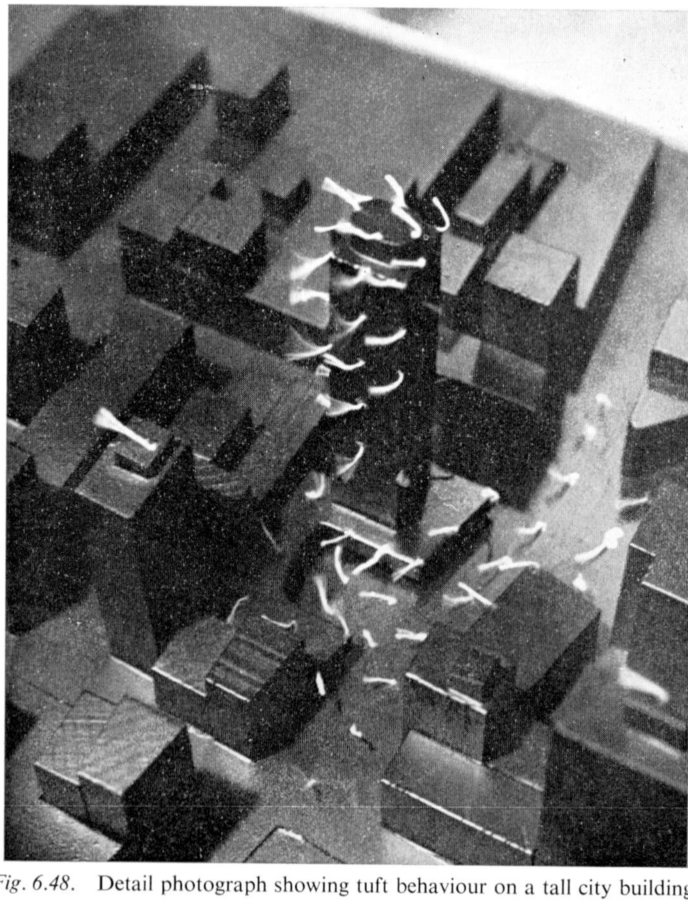

Fig. 6.48. Detail photograph showing tuft behaviour on a tall city building.

Fig. 6.49. Projection type water tunnel for two dimensional flow visualisation.

Where flow conditions can be reasonably assumed to be of a two dimensional nature such as airflow over the centre portion of a long low building of constant cross-section normal to the wind, or near

Fig. 6.50. A highly misleading 'potential flow' pattern lacking essential features such as separation, shear layers and broad turbulent wake.

the mid height of an isolated tall slender building of constant cross-section, a projection water tunnel such as in Fig. 6.49 can be used. This device is also useful in the study of internal airflow patterns provided changes in jet direction indicated in three dimensional studies are principally in one plane. Visualisation in this instrument is achieved by small styrene pellets in the stream.

Flow visualisation devices are available which produce flow patterns similar to those predicted by potential flow theory. Such flow patterns (Fig. 6.50) are extremely misleading as they lack the essential characteristics of airflow around buildings such as separation, shear layers and a broad turbulent wake.

Dynamic Viscosity and Density of Air and Water as a Function of Temperature

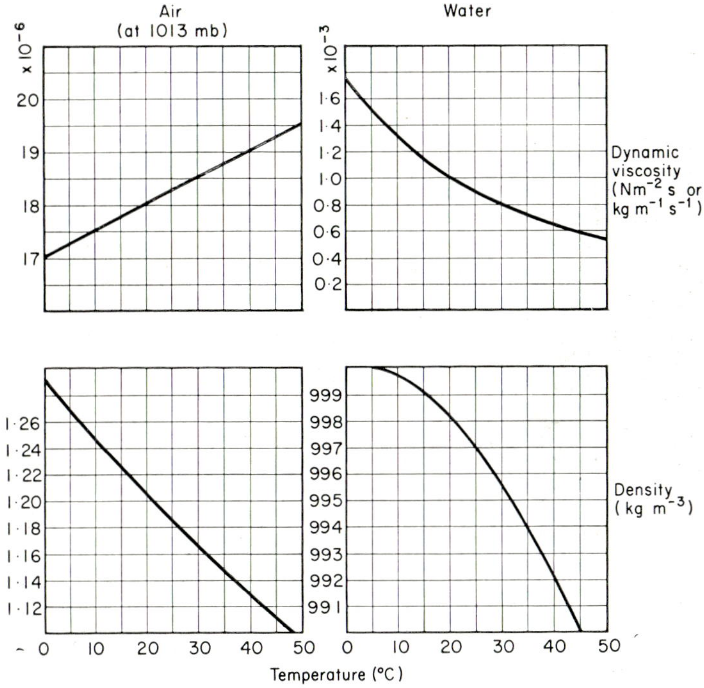

References

Chapter 1

1.1. Hilberseimer, L., *The Nature of Cities*, Paul Theobold & Co., Chicago, 1955.

1.2. Cotterell, A. and Morgan, D., *China; Integrated Study*, Harrap, London, 1975.

1.3. Brown, S., *World of the Wind*, Bobbs-Merrill Co., Indianapolis, 1961.

1.4. School of Civil Engineering, University of Sydney 'Post Graduate Course; The Structural and Environmental Effects of Wind on Buildings and Structures', Sydney, May, 1975.

1.5. Vitruvius, *The Ten Books of Architecture*, translated by Morris Hickey Morgan, Dover, New York, 1960.

1.6. Wittkower, R., *Architectural Principles in the Age of Humanism*, 2nd edn., Alec Tiranti Ltd., London, 1952.

1.7. Violich, F., *Cities of Latin America; Housing and Planning to the South*, Reinhold, New York, 1944.

1.8. Harada, J., *The Lessons of Japanese Architecture*, Studio Vista, London, 1936.

1.9. Miner, H., *St. Denis, a French-Canadian Parish*, University of Chicago, Chicago, 1939.

1.10. Olgyay, V., *Design with Climate*, Princeton University Press, Princeton, N.J., 1963.

1.11. Aronin, J. E., 'Climate and Architecture', Reinhold, New York, 1953.

1.12. Kampffmeyer, H., *Homes Should be Built near Workshops*, Julius Hofman Verlag, Stuttgart, 1932.

1.13. Mayer, A., 'The New Capital of the Punjab,' *Journal of the American Institute of Architects*, Oct. 1950, p. 168.

1.14. Bender, C. B., 'The design of structures to resist wind pressures', *Proceedings of the Institution of Civil Enigneers*, Lond., Vol. 69, 1882, pp. 80–119 (Paper 1770).

1.15. George, W. K. and Lumley, J. L., 'The laser velocimeter and its applications to the measurement of turbulence', *Journal of Fluid Mechanics*, **69**, 1973, pp. 321–362.

1.16. Gossard, E. E. *et al.*, 'Middle scale wind patterns revealed by dual-doppler radar', Paper I-4, *2nd U.S. National Conference on Wind Engineering Research*, 22–25 June 1975, sponsored by the Wind Engineering Research Council, Colorado State University, Fort Collins, 1975.

1.17. Smeaton, J., 'An experimental enquiry concerning the natural power of the wind to turn mills and other machines depending on circular motion', *Proceedings of the Royal Society II*, 1759, 366 pp.

1.18. Royal Navy, Admiralty Memorandum, 1838.

1.19. Westhofen, W., 'The Forth Bridge', *Engineering*, **49**, 28th Feb. 1890.

1.20. Ower, E. and Pankhurst, R. E., *Measurement of Air Flow*, 4th edn., Pergamon Press, Oxford, 1966.

1.21. Rouse, H. and Ince, S., *History of Hydraulics*, Dover, New York, 1963.

1.22. Streeter, V. L., *Fluid Mechanics*, 4th edn., McGraw-Hill, New York, 1966.

1.23. Stokes, G. G., 'On the theories of the internal friction of fluids in motion', *Transactions of the Cambridge Philosophical Society*, **8**, 1845.

1.24. Reynolds, O., 'An experimental investigation of the circumstances which determine whether the motion of water will be direct or sinuous of the law of resistance in parallel channels', *Phil. Trans. of the Royal Society*, London, 1883.

1.25. Prandtl, L. and Tietjens, O. G., *Fundamentals of Hydro- and Aeromechanics*, based on lectures of L. Prandtl by O. G. Tietjens, translated by L. Rosenhead, Dover, New York, 1957.

1.26. Von Karman and Rubach, 'On the Mechanism of Fluid Resistance', *Physik Z.*, 1912, p. 49.

1.27. Shaw, W. N., *Air Currents and the Laws of Ventilation*. University Press, Cambridge, 1907.

1.28. Series of A.S.H.V.E. Research Reports:

In *A.S.H.V.E. Trans.*, **30**, 1924, 'Air Leakage Through the Openings in Buildings' by F. C. Houghten and C. C. Schrader.

'Air Leakage around Window Openings' by C. C. Schrader.

In *A.S.H.V.E. Trans.*, **34**, 1928, 'Air Leakage Studies on Metal Windows in a Modern Office Building', by F. C. Houghten and M. E. O'Connell.

'Air Leakage Through a Pivoted Metal Window' by F. C. Houghten and M. E. O'Connell.

'Effect of Frame Calking and Storm Windows on Infiltration Around and Through Windows' by W. M. Richtmann and C. Braatz.

'The Weather-tightness of Rolled Section Steel Windows' by J. E. Emswiler and W. C. Randall.

In *A.S.H.V.E. Trans.*, **37**, 1931, 'Air Infiltration Through Double-hung Wood Windows' by G. L. Larson *et al.*

In *A.S.H.V.E. Trans.*, **39**, 1933, 'Air Infiltration Through Steel Framed Windows' by D. O. Rusk *et al.*

In *A.S.H.V.E. Trans.*, **71**, 1965, 'Air Leakage Values for Residential Windows' by J. R. Sasaki and A. G. Wilson.

1.29. Marin, A., 'Influence of stack effect on the heat loss in tall buildings', *A.S.H.V.E. Transactions*, **40**, 1934, p. 377.

1.30. Dick, J. B., 'The fundamentals of natural ventilation of houses', *Journal of the Institution of Heating and Ventilating Engineers*, June 1950, pp. 123–134.

1.31. Hidy, G. M., *The Winds; the Origins and Behaviour of Atmospheric Motion*, Van Nostrand, Princeton, 1967.

1.32. Report of the Court of Inquiry on the Tay Bridge Disaster on 28th Dec. 1879. London, 1880.

1.33. Stoney, B. B., *The Theory of Stresses in Girders and Similar Structures with Practical Observations on the Strength and Other Properties of Materials*, new edn., Longmans, Green and Co., London, 1886.

1.34. Kernot, W. C., 'Wind Pressures', *Australasian Builder and Contractors News*, **13**, 21 Oct. 1893, p. 194.

1.35. Irminger, J. O. V., 'Nogle Forsog over Trykforholdene paa Planer og Legemer paavirkede of Luftstromninger', *Ingenioren*, 1894 and *Engineering News and Engineers*, 1895.

1.36. Irminger, J. O. V. and Nokkentved, Chr., 'Wind-pressure on buildings; experimental researches (2nd series)', *Ingeniørvidenskabelige Skrifter*, A, No. 42, 1936.
1.37. Bailey, A., *Wind Pressure on Buildings*, Institution of Civil Engineers, Selected Engineering Papers No. 139, London, 1933.
1.38. Bailey, A. and Vincent, N. D. G., 'Wind pressures on buildings, including effects of adjacent buildings', *Journal of the Institution of Civil Engineers*, **20**, 1943, pp. 243–275.
1.39. Jensen, M., *Shelter Effect Investigations into the Aerodynamics of Shelter and its Effects on Climate and Crops*, Danish Technical Press, Copenhagen, 1954.
1.40. Sato, K. *et al.*, *Studies on Windbreaks*, Nippon-Gakujutsu-Shiukakai, Tokyo, 1932.
1.41. White, R. F., *Effects of Landscape Development on the Natural Ventilation of Buildings and their Adjacent Areas*, Texas Engineering Experiment Station, College Station, Research Report No. 45, 1954.
1.42. Smith, E. G., *The Feasibility of Using Models for Predetermining Natural Ventilation*, Texas Engineering Experiment Station, College Station, Research Report No. 26, 1951.
1.43. Caudill, W. W., *Geometry of Classrooms as Related to Natural Lighting and Natural Ventilation*, Texas Engineering Experiment Station, College Station, Research Report No. 36, July 1952.
1.44. Van Straaten, J. F. *et al.*, *Ventilation and Thermal Considerations in School Building Design*, C.S.I.R. Research Report, Pretoria, No. 203, 1965.
1.45. Van Straaten, J. F., *Thermal Performance of Buildings*, Elsevier, Amsterdam, 1967.
1.46. Jensen, M., *The Model-law for Phenomena in Natural Wind*, Ingenioren-International Ed., Vol. 2, No. 4, 1958, pp. 121–158.
1.47. Baines, W. Douglas, 'Effect of velocity distribution on wind loads and flow patterns on buildings', Paper from *Proceedings of International Conference on Wind Effects on Buildings and Structures*, held at N.P.L., 26–28 June 1963, Teddington, N.P.L., 1963.
1.48. Hamilton, G. F., *Effect of Velocity Distribution on Wind Loads on Walls and Low Buildings*, University of Toronto Technical Paper No. 6205, Toronto, Nov. 1962.
1.49. Trimble, T. H., *A Wind Tunnel Investigation of the Airflow Around a Tall City Building*, Aerodynamics Note No. 238, Aeronautical Research Laboratories, Melbourne, Jan. 1965.
1.50. Davenport, A. G., *A Report on the Activities of the Boundary Layer Wind Tunnel Laboratory, November 1965 to March 1969*, Faculty of Engineering Science, The University of Western Ontario, London, Canada, March 1969.
1.51. Wise, A. F. E., *Wind Effects Due to Groups of Buildings*, B.R.S. Current Paper 23, 1970.
1.52. Cermak, H. E. *et al.*, *Investigation of the Candlestick Park Wind Problem*, Vol. II, Wind Tunnel Model Study Technical Report, Metronics Associates, Palo Alto, Cal., July 1963.
1.53. Brisker, S. H. and O'Hare, M., 'With wind tunnels design is a breeze', *Progressive Architecture*, March 1967, pp. 166–171.
1.54. O'Hare, M. and Kronauer, R. E., 'Fence designs to keep winds from being a nuisance', *Architectural Record*, **146**, July 1969, pp. 151–156.

1.55. Newberry, C. W. and Wise, A. F. E., 'How wind shapes buildings; wind effects on buildings and structures', *Discovery*, **24**, July 1962, pp. 28–34.

1.56. Dalgliesh, W. A. *et al.*, 'Wind pressure measurements on a full-scale high-rise office building', Paper from *Proceedings of International Research Seminar on Buildings and Structures*, held in Ottawa, September 1967, pp. 167–200. Ottawa, National Research Council, Oct. 1968 (Division of Building Research, Research Paper 379).

1.57. Marshall, R. D., *A Study of Wind Pressures on a Single-family Dwelling in Model and Full Scale*, N.B.S. Technical Note 852, National Bureau of Standards, Washington, Oct. 1974.

1.58. Joubert, P. N. *et al.*, 'The drag of bluff bodies immersed in a turbulent boundary layer', Paper from *International Research Seminar on Buildings and Structures*, held in Ottawa, Sept. 1967, pp. 297–335. Ottawa, National Research Council, Oct. 1968.

1.59. Fricke, F. R., 'Noise inside buildings due to external airflows', *Applied Acoustics*, **5**, 1972, 233–235.

1.60. 'Testing houses for leaks', *B.R.E. News*, No. 33, Autumn 1975.

1.61. Vickery, B. J. and Apperley, C. W., *On the Prediction and Evaluation of the Ground Wind Environment*, University of Sydney, Department of Civil Engineering Research Report R227, Sydney, Nov. 1973.

1.62. Melbourne, W. H. and Joubert, P. N., 'Problems of wind flow at the base of tall buildings', *International Conference on Wind Effects on Buildings and Structures*, held at Science Council of Japan, Tokyo, 6–9 Sept. 1971. Tokyo Saikon Shuppan Co. Ltd., 1971.

1.63. Aynsley, Richard M., 'The environment around tall buildings', *Proceedings of the Australian and New Zealand Conference on the Planning and Design of Tall Buildings*, Sydney, 14–17 Aug. 1973. Joint Committee Tall Buildings, Lehigh University, Bethlehem, June 1974, pp. 256–274.

1.64. *International Conference on Wind Effects on Buildings and Structures*, held at the N.P.L., 26–28 June 1963, Teddington, N.P.L., 1963.

1.65. *International Seminar on Wind Effects on Buildings and Structures*, held at the National Research Council of Canada, 11–15 Sept. 1967, 2 Vols. Ottawa, N.R.C., 1967.

1.66. *International Conference on Wind Effects on Buildings and Structures*, held at Science Council of Japan, Tokyo, 6–9 Sept., 1971, Tokyo, Saikon Shuppan Co. Ltd., 1971.

1.67. *Fourth International Conference on Wind Effects on Buildings and Structures*, 8–12 Sept., 1975, Heathrow Hotel, London, Cambridge University Press, 1976.

1.68. Eaton, K. J. and Mayne, J. R., 'The measurement of wind pressure on two-storey houses at Aylesbury', *Journal of Industrial Aerodynamics*, **1**, 1975, pp. 67–109.

Chapter 2

2.1. British Standards, *Methods for the Measurement of Fluid Flow in Pipes*, B.S. 1042, Part 1, 1964 (alt. A.S.M.E. Fluid Meters Report, Part 1, p. 100).

2.2. Salter, C., Warsop, J. H. and Goodman, D. G. ,*A New Design of Pitot-static Tube with a Discussion of Pitot-static Tubes and their Calibration Factors*, N.P.L., Aero Dept. 1013, 1962 (also ARC 23, 760).

2.3. Van Straaten, J. F., *Thermal Performance of Buildings*, Architectural Science Series, Elsevier, 1967.
2.4. Ward Smith, A. J., *Pressure Losses in Ducted Flows*, Butterworths, 1971.
2.5. Langhaar, H. C., *Dimensional Analysis and Theory of Models*, John Wiley (New York), Chapman and Hall (London), 1951.
2.6. Bearman, P. W., 'Some recent measurements on the flow around bluff bodies in smooth and turbulent flow', *Symposium on External Flows*, Bristol University 1972.
2.7. Vickery, B. J., 'Fluctuating lift and drag on a large cylinder of square cross-section in a smooth and turbulent stream', *Journal of Fluid Mechanics*, **25**, pp. 481–494, 1966.
2.8. Gartshore, I. S., *The Effects of Freestream Turbulence on the Drag of Rectangular Two Dimensional Prisms*, University of Western Ontario, Canada, BLWT-4-1973.
2.9. Fage and Falkner, *Further Experiments on Flow Around a Circular Cylinder*, ARC RM 1369, 1931.
2.10. Delaney, N. K. and Sorenson, N. E., *Low Speed Drag of Cylinders of Various Shapes*, NACA TN 3038, 1953.
2.11. Fage and Warsap, *Effects of Turbulence and Surface Roughness on Drag of Circular Cylinders*, ARC RM 1283, 1929.

General reference texts on fluid mechanics:

Bradshaw, P., *Experimental Fluid Mechanics*, Pergamon Press, 1964.
Bradshaw, P., *An Introduction to Turbulence and its Measurement*, Pergamon Press, 1971.
Duncan, W. J., Thom, A. S. and Young, A. P., *Mechanics of Fluids*, Edward Arnold, 2nd edn., 1970.
Shames, I. H., *Mechanics of Fluids*, McGraw-Hill, 1962.
Streeter, V. L., *Fluid Mechanics*, McGraw-Hill, 4th edn., 1966.

Chapter 3

3.1. Whittingham, H. E., *Extreme Wind Gusts in Australia*, Bulletin No. 46, Bureau of Meteorology, Melbourne, Australia, Feb. 1964.
3.2. Thom, H. S. C., 'Tornado probabilities', *Monthly Weather Review*, pp. 730–736, Oct./Dec. 1963.
3.3. Mehta, K. C., Minor, J. E. *et al.*, *Engineering Aspects of the Tornadoes of April 3–4, 1974*, Committee on Natural Disasters, National Academy of Sciences, Washington, D.C., 1975.
3.4. Deacon, E. L., 'Gust variation with height up to 150 m', *Quart. J. Roy. Met. Soc.*, **81**, No. 350, Oct. 1955, pp. 562–573.
3.5. Gomes, L. and Vickery, B. J., 'On the prediction of extreme wind speeds from the parent population', *Int. J. Industrial Aerodynamics*, **2**, 1976 (in press).
3.6. Van der Hoven, I., 'Power spectrum of horizontal wind speed in the frequency range from 0·0007 to 900 cycles/hr', *J. Meteor.*, **14**, p. 160, 1957.
3.7. Davenport, A. G., 'The spectrum of horizontal gustiness near the ground in high winds', *Quart. J. Roy. Met. Soc.*, **87**, p. 194, 1961.
3.8. *SAA Loading Code, Part 2—Wind Forces*, Australian Standard AS1170. Standards Association of Australia, Sydney, 1975.

3.9. De Bray, B. G., *Shear Flow Over Ramps or Escarpments*, Interim Report, Faculty of Engineering, University of Auckland, N.Z., 1972.

3.10. Gumbel, E. J., *Statistics of Extremes*, Columbia University Press, New York, 1958.

Further Reading

Lumley, J. L. and Panofsky, H. A., *The Structure of Atmospheric Turbulence*, Interscience Publishers, New York, 1964.

Macdonald, A. J., *Wind Loading on Buildings*, Applied Science Publishers, London, 1975.

Riehl, H., *Introduction to the Atmosphere*, McGraw-Hill, New York, 1965.

Chapter 4

4.1. Macdonald, A. J., *Wind Loading on Buildings*, Applied Science Publishers, London, 1975.

4.2. Warburton, G. B., *The Dynamical Behaviour of Structures*, Pergamon Press, Oxford, 1964.

4.3. Robson, J. D., *An Introduction to Random Vibration*, Edinburgh University Press, Edinburgh, 1964.

4.4. Crandall, S. H. and Mark, W. D., *Random Vibrations in Mechanical Systems*, Academic Press, New York, 1963.

4.5. *S.A.A. Loading Code, Part 2—Wind Force*, AS 1170, Standards Association of Australia, North Sydney, 1975.

4.6. Eaton, K. J. and Mayne, J. R., 'The measurement of wind pressures on two-storey houses at Aylesbury', *Journal of Industrial Aerodynamics*, **1**, No. 1, June 1975.

4.7. Davenport, A. G.,'Gust loading factors', *Journal Struct. Div., Proc. A.S.C.E.*, **93**, June 1967.

4.8. Vickery, B. J., 'On the assessment of wind effects on elastic structures', *C.E. Trans. Inst. Eng. Aust.*, Oct. 1966, pp. 183–192.

4.9. Vickery, B. J., 'On the reliability of gust loading factors', *Civil Eng. Trans. Inst. Eng. Aust.*, April 1971, pp. 1–9.

4.10. Simiu, E., 'Gust factors and alongwind pressure correlations', *J. Struct. Div. Proc. A.S.C.E.*, **100**, No. ST9, Sept. 1974.

4.11. Simiu, E., 'Wind spectra and dynamic alongwind response', *J. Struct. Div. Proc. A.S.C.E.*, **100**, No. ST9, Sept. 1974.

4.12. Dalgleish, W. A., 'Comparison of model/full-scale wind pressures on a high rise building', *J. Industrial Aerodynamics*, **1**, No. 1, June 1975.

4.13. Vickery, B. J. and Clark, A. W., 'Lift or across-wind response of tapered stacks', *J. Struct. Div., Proc. A.S.C.E.*, **98**, No. ST1, Jan. 1972.

4.14. Vickery, B. J., 'The provisions and limitations of the Australian wind loading code, AS1170—Part 2', *Proc. Aust. Conf. on Planning and Design of Tall Buildings*, University of Sydney, Aug. 1972.

4.15. Melbourne, W. H., 'Cross-wind response of structures to wind action', *Proc. 4th Int. Conf. on Wind Effects on Buildings and Structures*, London, 1975.

4.16. Davenport, A. G., 'Approaches to the design of tall buildings against wind', Theme Paper, Committee 7, *ASCE-IABSE Conf. on Planning and Design of Tall Buildings*, Lehigh University, 1972.

4.17. Davenport, A. G., 'Design criteria for wind action', *The Structural and Environmental Effects of Wind on Buildings and Structures*, Chapt. 8, University of Sydney, 1975.
4.18. Chen, P. and Robertson, L. E., 'Human perception of motion', *J. Struct. Div., Proc. A.S.C.E.*, **99**, 1973.
4.19. Robertson, L. E., 'Limitations of swaying motion of tall buildings imposed by human response factors', *Proc. Aust. Conf. on Planning and Design of Tall Buildings*, University of Sydney, 1972.
4.20. Hansen, R. J., Reed, J. W. and Vanmarke, E., 'Human response to wind induced motion of buildings', *J. Struct. Div., Proc. A.S.C.E.*, **99**, 1973.

Chapter 5

5.1a. Melbourne, W. H., *Ground Level Winds Caused by Large Buildings*, Monash University, Dept. of Mechanical Engineering, MMER 4, 1971.
5.1b. Melbourne, W. H. and Joubert, P. N., 'Problem of wind flow at the base of tall buildings', *Proc. Int. Conf. on Wind Effects on Buildings and Structures*, Tokyo, 1971, pp. 105–114.
5.2a. Davenport, A. G., 'An approach to human comfort criteria for environmental wind conditions', *CIBA/WMO Colloquium Teaching the Teachers 1972*, Swedish National Building Research Institute, Stockholm.
5.2b. Davenport, A. G., 'Approaches to the design of tall buildings against wind', *Int. Conf. Planning and Design of Tall Buildings*, Lehigh University, Pennsylvania, 1972, pp. 1–22.
5.3. Wise, A. F. E., 'Wind effects due to groups of buildings', *Phil. Trans. of Royal Society London*, **A269**, 1971, pp. 469–485.
5.4a. Isyumov, N. and Davenport, A. G., 'The ground level wind environment in built up areas', *Proc. 4th Int. Conf. on Wind Effects on Buildings and Structures*, London, 1975, Cambridge University Press.
5.4b. Isyumov, N. and Davenport, A. G., 'Comparison of full scale and wind tunnel wind speed measurements in the Commerce Court Plaza', *Jnl of Industrial Aerodynamics*, **1**, pp. 201–212, 1975.
5.5. Melbourne, W. H., 'Full scale measurements of wind effects projects at Monash University, Australia', *Int. Symposium on Full Scale Measurements of Wind Effects on Tall Buildings and Other Structures*, University of Western Ontario, June 1974.
5.6. Penwarden, A. D., 'Acceptable wind speeds in towns', *Building Science*, **8**, 1973, pp. 259–267.
5.7a. Lawson, T. V. and Penwarden, A. D., 'The effects of wind on people in the vicinity of buildings', *Proc. 4th Int. Conf. on Wind Effects on Buildings and Structures*, London, 1975, Cambridge University Press.
5.7b. Lawson, T. V., *The Wind Environment of Buildings: a Logical Approach to the Establishment of Criteria*, University of Bristol, Dept. of Aero Eng. Report TVL7321, 1973.
5.8. Penwarden, A. D. and Wise, A. F. E., *Wind Environment Around Buildings*, Building Research Establishment, England, Report, H.M.S.O., 1975.
5.9. Wardlaw, R. L. and Moss, G. F., *A Standard Tall Building Model for the Comparison of Simulated Natural Winds in Wind Tunnels*, CAARC Report CC662, Tech, 25, Jan. 1970.

5.10. Vickery, B. J., Contribution to Course Notes, *The Structural and Environmental Effects of Wind on Buildings and Structures*, University of Sydney and Monash University, 1975.
5.11. Melbourne, W. H., 'Wind tunnel modelling of buoyant chimney plume', *Proc. 3rd Australasian Conf. on Hydraulics and Fluid Mechanics*, Institution of Engineers, Australia, Nov. 1968.
5.12. Melbourne, W. H. and Gartshore, I.S., A wind tunnel model study of air conditioning contamination by intake of flue gas. *Int. Conference on Instrumentation in Aerospace Simulation Facilities.* IEEE Publication 75, CHO 993-6, AES.

Chapter 6

6.1. Consolazio, W. V. and Pecora, L. J., 'Minimal replenishment air required for living spaces', *A.S.H.V.E. Transactions*, **53**, 1947, p. 127.
6.2. Givoni, B., *Man, Climate and Architecture*, 2nd edn., Applied Science, London, 1976.
6.3. Macfarlane, W. V., 'Thermal comfort zones', *Architectural Science Review*, **1**, No. 1, Nov. 1958, pp. 1–14.
6.4. Jackman, P. J., 'A study of the natural ventilation of tall office buildings', *Heating and Ventilating Engineer and Journal of Air Conditioning*, **44**, No. 519, Oct. 1970, pp. 183–195.
6.5. Sexton, D. E., *Building Aerodynamics*, Building Research Station, Garston, England, Current Paper No. 64/68, Aug. 1968.
6.6. Koenigsberger, O. H. *et al.*, *Manual of Tropical Housing and Building*, Part I, *Climatic Design*, Longman Group Limited, London, 1973.
6.7. Fry, A. and Drew, J., *Tropical Architecture*, B. T. Bashford Ltd., London, 1964.
6.8. Olgyay, V., *Design with Climate; Bioclimatic Approach to Architectural Regionalism*, Princeton University Press, Princeton, N.J., 1963.
6.9. Van Straaten *et al.*, *Ventilation and Thermal Considerations in School Building Design*, C.S.I.R. Research Report 203, C.S.I.R., Pretoria, 1965.
6.10. Caudill, W. W. and Reed, Bob H., *Geometry of Classrooms as Related to Natural Lighting and Natural Ventilation*, Research Report No. 36, Texas Engineering Experiment Station, College Station, Texas, July 1952.
6.11. Baturin, V. V., *Fundamentals of Industrial Ventilation*, Pergamon Press, Oxford, 1972.
6.12. Weston, E. T., *Natural Ventilation in Industrial-type Buildings*, Special Report No. 14, Commonwealth Experimental Building Station, Sydney, Feb. 1954.
6.13. Givoni, B., *Basic Study of Ventilation Problems in Housing in Hot Countries*, Report of Building Research Station, Technion, Technion City, Oct. 1962.
6.14. *A.S.H.R.A.E. Guide and Data Book; Systems and Equipment for 1967*, A.S.H.R.A.E., N.Y., 1967.
6.15. Van Straaten, J. F., *Thermal Performance of Buildings*, Elsevier, Amsterdam, 1967.
6.16. Chien, Ning *et al.*, *Wind-tunnel Studies of Pressure Distribution on Elementary Building Forms*, Iowa Institute of Hydraulic Research, State University of Iowa, Iowa City, 1951.

6.17. Jensen, M. and Franck, N., *Model-scale Tests in Turbulent Wind*, Part II, *Phenomena Dependent on the Velocity Pressure. Wind Loads on Buildings*, The Danish Technical Press, Copenhagen, 1965.
6.18. *SAA Loading Code Part 2—Wind Forces*, Australian Standard 1170, Part 2, 1973.
6.19. British Standard Code of Practice, *Code of Basic Data for the Design of Buildings*, CP3, Chapter V, Loading (Part 2, Windloads), 1972.
6.20. Aynsley, R. M., *A Study of Airflow Through and Around Buildings*, Ph.D. Thesis, University of New South Wales, School of Building, 1976.
6.21. Bailey, A. and Vincent, N. D. G., 'Wind pressures on buildings, including effects of adjacent buildings', *Journal of the Institution of Civil Engineers*, **20**, 1943, pp. 243–275.
6.22. Drysdale, J. W., *Designing Houses for Australian Climates*, Commonwealth Experimental Building Station, Sydney, 1975.
6.23. *Climate of Australia*, Extract from Official Year Book of Australia, No. 59, 1973, Dept. of Science, Bureau of Meteorology, Melbourne, 1973.
6.24. Dick, J. B., 'The fundamentals of natural ventilation of houses', *Journal of the Institution of Heating and Ventilating Engineers*, **18**, No. 179, 1950.
6.25. Wannenburg, J. J. and Van Straaten, J. F., 'Wind tunnel tests on scale model buildings as a means for studying ventilation and allied problems', *J. Institution of Heating and Ventilating Engineers*, March 1957.
6.26. Snyckers, W. A., *Wind Tunnel Studies of the Flow of Air Through Rectangular Openings with Application to Natural Ventilation of Buildings*, Master of Science in Engineering Thesis, University of Pretoria, Dec. 1970.
6.27. Caudill, W., *Some General Considerations in the Natural Ventilation of Buildings*, Research Report No. 22, Texas Engineering Experiment Station, College Station, Feb. 1951.
6.28. Holleman, T., *Air Flow Through Conventional Window Openings*, Research Report No. 33, Texas Enginering Experiment Station, College Station, Nov. 1951.
6.29. Weston, E. T., *Air Movement in Industrial Buildings; Effects of Nearby Buildings*, Special Report No. 19, Commonwealth Experimental Building Station, Sydney, 1956.
6.30. Chand, I., *Prediction of Air Movement in Buildings*, Building Digest No. 100, Central Building Research Institute, Roorkee, India, Sept. 1972.
6.31. Smith, E. G., *The Feasibility of Using Models for Predetermining Natural Ventilation*, Research Report No. 26, Texas Engineering Experiment Station, College Station, Texas, 1951.
6.32. McKeon, R. J. and Melbourne, W. H., 'Wind tunnel blockage effects and drag on bluff bodies in a rough wall boundary layer,' Paper from the *Proceedings of the Third International Conference on Wind Effects on Buildings and Structures*, held at Science Council of Japan, Tokyo, Sept. 6–9, 1971, Saikon Shuppan Co. Ltd., Tokyo, 1971.
6.33. Daws, L. F., 'Movement of air streams indoors', *Journal of the Institution of Heating and Ventilating Engineers*, **37**, Feb. 1970, pp. 241–253.
6.34. White, R. F., *Effects of Landscape Development on the Natural Ventilation of Buildings and their Adjacent Areas*, Research Report No. 45, Texas Engineering Experiment Station, College Station, March 1954.
6.35. Geiger, R., *The Climate Near the Ground*, translated from 4th German edn. Cambridge, Mass., Harvard University Press, 1966.

Index

After-body wake length, 70
Airflow through openings
  depth of reveals, 204
  discharge coefficients, 188
    sources of, 202
  free area of cracks, 186
  jets, 204
  openings in series, 192
  rectangular openings, 202
  single openings, 188
  spacing of openings, 204
  unequal openings, 185, 224
Alberti, Leon Battista, 6
Anemometers
  anemograph, 80, 81, 84
  cup, 17
  Dines, 17, 20
  Doppler, 12
  hot wire or film, 18, 77, 78, 164
    advantages, 164
    effects of turbulence on, 164
  mounting height, 17
  pendulum, 13, 23
  pitot-static tube, 23, 48, 49
    disadvantage, 166
  pressure plate, 13
  thermocouple, 19
  see also Wind speed measurement
Architectural features, 195
  arcades, 154, 169, 175
  aspect ratio, 154
  average wall porosities, 210
  canopies, 173, 174
  colonnades, 175
  extended eaves, 201
  high set homes, 212
  low set homes, 225
  podiums, 174
Aristotle, 2
Athens, 4
Australian Wind Code, 98

Bailey, A., 32, 35, 195
Baines, W. D., 35
Baker, B., 17, 23
Barometer, 12, 14, 29
Baturin, V., 188
Bearman, P. W., 68, 70
Beaufort, F., 17
Bernoulli, D., 26
  equation, 43, 45, 48, 50, 69
  principle, 26
Bluff bodies, 58
Bouguer, P., 13
Boundary layers, 48, 62
  earth's, 85
  false ground level, 90
  full scale wind speed profile measurement, 94
  gradient height, 89, 104, 167
  logarithmic law, 89, 90
  power law, 89
Brisbane, 200
Buckingham Π theorem, 59
Buildings
  CAARC, 166
  circular, 174
  exposure to wind, 173
  groups, 195
  Japanese houses, 7
  low rise, 128
  orientation, 2, 4, 7, 8, 184, 200, 209
  rectangular, 174
  relative placement of, 173

Buildings—*contd.*
schools, 35
tall, 28, 39, 128, 143, 184: *see also* Dynamics of structural systems
unusual, 225

Cauchy number, 60
Caudill, W., 188
Ceiling fan, 232
Cermak, H. E., 39
Chien, N., 194
Clark, A. W., 143
Coefficients, 60
correction coefficient, 49
discharge coefficient, 188, 202
drag coefficient, 67, 70
dynamic loss coefficient, 192
loss coefficient, 56
moment coefficient, 60
pressure coefficient or ratio, 168, 189
reference height corrections, 190
velocity coefficient or ratio, 154, 160
Comfort, *see* Human thermal comfort
Computers, 25, 78
Continuity, principle of, 45, 48
Control volume, 52
Crandall, S. H., 114
Cyclones, *see* Wind climate

Davenport, A. G., 147, 149, 156 159, 160, 162
universal spectrum, 97
da Vinci, L., 12, 25
De Bray, B. G., 99
Delaney, N. K., 67, 70
Dick, J. B., 28, 202
Dimensional analysis, 58
Drag
falling model method, 23

Drag—*contd.*
friction, 58
pressure, 58
Dynamics of structural systems, 108
accelerations, 148
aerodynamic damping, 143
negative, 145
broad band excitation, 143
cross-wind oscillations of tall structures, 71, 142, 143
damping constant, 109
deflections, 148
dynamic amplification factor, 112, 120, 142
forced vibration, 111
free vibration, 110
guy wires, 145
logarithmic decrement, 110
natural frequency, 110, 120, 142
resonant peak displacement, 113
spring–mass–dashpot system, 108
spring stiffness, 109
tall welded stacks, 143
transmission wires, 145

Ellis, F. P., 181
Entrainment, 70
Environmental wind problems
near corners, 154, 175
occupant sensitivity quotient, 149
pedestrian wind problems, 35, 38
standing vortex, 151, 172, 174
see also Pollution
Euler, L., 26

Ferdinand II, 29
Flow
freestream, 43, 48
incompressible, 46
irrotational, 46
laminar, 63
separation, 62, 65
transition, 63

Flow—*contd.*
visualisation, 25, 43, 152, 220, 232
potential flow devices, 236
smoke, 232
tufts, 43, 233, 234
water tunnel, 236
Forth Bridge, 17
Franck, N., 195
Franklin, B., 30
Froude number, 60
Fry, A., 188
Full scale wind studies, 39

Galileo, G., 12, 26
Gartshore, I. S., 69, 70
Gas law, 188
Geiger, R., 231
Givoni, B., 181, 188, 189
Greek wind names, 4
Gust
energy, 98
factor, 124
size, 98
wavelength, 98
wind speed, 98

Hadley, G., 30
Halley, E., 29
Hamilton, G. F., 35
Helmholtz, H., 27
Hoerner, S. F., 66, 67
Holleman, T., 231
Hooke, R., 13, 15
Human thermal comfort, 181
airflow to restore, 183
estimates, 215, 227
evaporative cooling, 181
external shading, 182
indoor, 180
thermal comfort zone, 181
Hurricanes, *see* Wind climate

Insect screens, 231
Irminger, J., 32
Isymov, N., 159, 160

Jackman, P. J., 188
Jensen, M., 34, 35, 195
Jesuit priests, 29
Jets, 204
Joubert, P., 39, 159

Kelvin, Lord W., 27
Kernot, W., 32
Koenigsberger, O., 188

Laminar flow, 63

Macfarlane, W., 181
Mach number, 60
Mark, W., 114
Mass spectrometer, 179
Melbourne, W., 153, 154, 157, 159, 160, 176, 179
'Meteorologica', 2
Momentum transfer, 57, 64
Moss, G., 166
Mulguf, 9

Natural ventilation, 35, 54
advantages and disadvantages of pressure difference method, 210
airflow through openings, 189, 192
cold climates, in, 184
combined effects of stack and wind pressure, 188
effects of landscaping, 231
energy sources for natural ventilation, 182

Natural ventilation—*contd.*
- infiltration, 28, 184
- internal airflow patterns, 229
- laws of, 28
- stack effect, 54
- temperate climates, in, 184
- use of louvres, 229
- wall effect, 204
- warm humid climates, in, 180, 181, 188, 195, 210
- wind scoops, 8, 9

Navier, C., 27
Navier–Stokes equations, 27
Newton, I., 26

O'Hare, M., 39
Olgyay, V., 11, 188
Openings
- effect on internal pressure, 133
- rectangular, 202
- series, in, 192
- single, 204
- spacing, 204
- unequal, 185, 224

Organisations
- Aeronautical Society of Great Britain, 31
- ASHRAE, 180, 202
- ASHVE, 28
- Building Research Station, England, 28, 39, 162
- Commonwealth Aeronautical Advisory Research Council, 166
- International Association for Wind Engineering, 40
- international conferences, 40
- International Meteorological Organisation, 29
- International Wind Study Group, 40
- ISHVE, 180
- National Physical Laboratory, 40, 51

Organisations—*contd.*
- National Research Council, Canada, 39
- Royal Society, 30
- Texas Engineering Experiment Station, 35, 224
- US National Bureau of Standards, 39

Osler, F., 13
Ower, E., 25

Pascal, B., 12, 26
- law, 26

Pedestrian wind speed criteria, 154, 156
Penwarden, A., 157, 159
Pitot, H., 13
- static tube, 23, 48, 49
- tube, 166

Poiseuille, J., 27
Pollution, 10
- air conditioning intakes, 179
- airborne pollutants, 176
- heights of chimney stacks, 176
- helium plume, 179
- helium tracer, 179
- initial plume momentum, 178
- location of stacks, 179
- plume buoyancy, 178
- plume mass flow rate, 178

Prandtl, L., 27, 63
Precautions for ground level winds, 173
Pressure
- atmospheric, 12, 29
- base or wake, 68
- differences, 182, 188
- dynamic (or velocity), 23, 119, 164
- external, distribution, 126
- head, 53
- internal, distribution, 133
- large local negative, effects, 141
- peak wind, 121

Pressure—*contd.*
  stagnation, 49
  static, 23, 46, 48, 164
  total, 23, 49

Real fluids, 42
Reed, B., 188
Reynolds, O., 27
  number, 27, 49, 51, 60, 62, 65
Risk of failure, 147
Robins, B., 30
Robinson, T., 13
Robson, J., 114
Rouse, H., 17, 30

Sailing vessels, 7
Saint-Venant, A., 27
Sato, K., 35
Shaw, W., 28
Shear layers, 43, 48, 68
  re-attachment, 70
  separation, 43
Size factor, 17, 120
Smeaton, J., 17, 21, 30
Snyckers, W., 202, 204
Sorensen, N., 67, 70
Stack effect, 182
  application of, 182, 184
  discharge due to, 185
  limitations of, 182
Statistical properties of wind
  mean wind speed, 86, 88
  modal wind speed, 105
  standard deviation of wind speed, 86
Stochastic process, 42
Stokes, G., 27
Streamlines, 43
Strouhal number, 141
Structural design criteria, 147
Sun control, 202
Système International (SI), 62

Tall buildings, 28, 39, 128, 143
'Ten Books on Architecture', 4
Theophratus, 3
Thermodynamics, first law of, 51
Topographic effects on wind, 98
  terrain categories, 118
  terrain roughness effects, 75, 85, 89
Tornadoes, *see* Wind climate
Torricelli, E., 12, 29
Tower of the Winds, 4, 5, 12
Town planning
  ancient towns, 1
  Buenos Aires, 7
  Canadian settlements, 8
  Chandigarh, 10
  Chinese, 12
    Chou dynasty, 2
  Egypt, 1, 9, 12
  'Feng-shui' doctrine, 2
  fortified towns, 4
  garden cities, 10
  Greek cities, 2
  healthy cities, 2, 5
  housing layouts, 2
  Hyderabad, 8, 9
  industrial pollution, 10
  Kahan, 1
  laws, 6, 10
  Letchworth, 10
  Magnetogorsk, 10
  New Jersey, 11
  Nuremburg, 10
  orientation of streets, 4
  Romans, 4
  shelter belts, 11, 35
  Spanish colonial towns, 6
  Sweden, 10
  Vienna, 10
  Welwyn, 10
Townsville, 201, 202, 212
Trimble, T., 39
Tropical cyclones, *see* Wind climate
Turbulence, 26, 40, 41, 42, 67

Turbulence—*contd.*
  intensity, 74, 92
Turbulent flow, 42, 70
Turbulent shear, 27

Unwin, W., 32

Van der Hoven, I., 96
Van Straaten, J., 35, 51, 194, 202, 204, 231
Velocity coefficients, 217
  advantages, 228
  disadvantages, 228
  reference wind speeds, 217
  sources, 223
Ventilation
  bathrooms, to, 180, 187
  bedrooms, to, 180, 187
  community halls, to, 180
  factories, to, 180
  kitchens, to, 180, 188
  living rooms, to, 180, 187
  minimum ventilation for health, 180, 187
  odour reduction, for, 181
  school classrooms, to, 180
  systems, 51, 52
    duct systems, 53
    energy equation, 51
    energy losses, 52
    fan pressure gain, 53
    flow rate through vents, 133
  tobacco smoke, for, 181
Venturi, G., 26
  meter, 48
Vickery, B. J., 68, 70, 143, 173
Vincent, N., 195
Viscosity, 42, 56, 58
  dynamic, 57
  law of, 26
Vitruvius, 4
Volumetric flow rate equation, 48, 192. 193
Von Karman, T., 27
Vortex
  generators, 65
  shedding, 28, 71, 142
Vortices, 43, 70

Wake flow, 43, 48
Wannenburg, J., 202
Wardlaw, R., 166
Wenham, F., 31
Weston, E., 188
Whirling arm, 30, 31
White, R., 35, 231
Wind characteristics
  altitude effects, 85
  Beaufort scale, 156, 157
  gradient wind speed, 89
  gust
    energy, 98
    size, 98
    speeds (peak speeds), 93
    wavelength, 98
  gustiness (turbulence), 86
  long term wind speed spectra, 95
  longitudinal velocity component, 42
  power spectral density, 74
  probability
    distribution of wind speed, 87, 94
    extreme wind speeds, of, 101, 104
  relation of peak gust to mean wind speed, 159
  return period, 106, 116
  sources of wind data, 195
  spectral density function, 95
  standard deviation of wind speed, 86, 87
  statistical characteristics, 86
  turbulent fluctuations, 92
  velocity or wind speed profile, 35, 41, 65, 89
  wind force scales, 17

Wind climate, 2, 5, 10
cold fronts, 79
cyclones, 79
dispersion wind speed, 105
extra-tropical cyclone, 80
global surface wind patterns, 79
line squalls, 79
mean wind speed, 86, 88
modal wind speed, 105
monsoon winds, 29
on-shore and off-shore winds, 85
origin of wind, 2, 79
thunderstorms, 79, 81
tornadoes, 79, 83
trade winds, 29
tropical cyclones, 82, 84
Weibull distribution, 101, 167
wind frequency data, 200
wind roses, 199
Wind damage
cladding failure, 148
risk, 147
roller doors failure, 139
structural failure, 147
Tay Bridge, 30
wind failures, 30
window failure, 139
Wind effects on heat transfer, 39
Wind generated noise, 39
Wind loading
across-wind forces, 141, 145
comparison of full scale and wind tunnel observations, 126
design wind speed, 116
drag force, 58, 144
gust factor methods, 124
advantages, 124
limit state design, 147
loads on internal walls or ceilings, 138
quasi-static wind loads, 72, 119, 142
roof loads, 128
size factor, 17, 120
torsional wind loads, 145
Wind loading—*contd.*
turbulence, effect of, 67, 70
ultimate loads, 116
vents and openings, effect of, 133
wind load spectrum, 120
wind loads from wind tunnel tests, 126
working loads, 116
Wind pressure measurement, 17, 32
electronic pressure transducers, 23, 166
manometer, 49, 50
multitube manometer, 23
pitot-static tube, 164
pressure coefficients, 121, 126, 194
Wind speed measurement, 12
balloon observations, 103, 166
fastest mile, 118
knots, 201
measurements, direct and indirect, 12
see also Anemometers
Wind tunnel testing, 71, 125, 143, 147, 150, 163, 166
data processing, 78
detail on models, 224
dynamic similarity, 59
fully dynamic tests, 72
model law, 35
model measurements, 76
modelling adjacent buildings, 163
roughness elements, 75, 76
scaling criteria, 60
simulated wind, 30
smoke diffusion studies, 72
strain gauged supports, 78
uniform flow, 32
vorticity generators, 76
Wind tunnels
aeronautical, 35, 71, 126
boundary layer, 32, 40, 75
close circuit, 74
early, 31, 41

Wind tunnels—*contd.*
modelling natural wind, 73, 160
Monash University, 74
open circuit, 76
University of Sydney, 75
University of Western Ontario, 198
working section, 74

Wise, A., 157

Xenophon, 3

Yaglou, C., 181